HIGH RELIABILITY
MANAGEMENT

HIGH RELIABILITY AND CRISIS MANAGEMENT
SERIES EDITORS Karlene H. Roberts and Ian I. Mitroff

HIGH RELIABILITY MANAGEMENT

Operating on the Edge

Emery Roe and Paul R. Schulman

STANFORD BUSINESS BOOKS
An imprint of Stanford University Press
Stanford, California

Stanford University Press
Stanford, California

Printed in the United States of America on acid-free, archival-quality paper

Library of Congress Cataloging-in-Publication Data

Roe, Emery.
  High reliability management : operating on the edge / Emery Roe and Paul R. Schulman.
      p. cm.—(High reliability and crisis management)
  Includes bibliographical references and index.
  ISBN 978-0-8047-5882-6 (cloth : alk. paper)—ISBN 978-0-8047-5946-5 (pbk. : alk. paper)
    1. Infrastructure (Economics)—Management. 2. Risk management.
3. Reliability (Engineering) 4. Electric utilities—California—Management—Case studies. 5. California ISO—Management. I. Schulman, Paul R. II. Title.
  HC79.C3R64 2008
  333.793′20685—dc22   2008001988

Designed by Bruce Lundquist
Typeset by Classic Typography in 10.5/15 Adobe Garamond

Special discounts for bulk quantities of Stanford Business Books are available to corporations, professional associations, and other organizations. For details and discount information, contact the special sales department of Stanford University Press. Tel: (650) 736-1783; Fax: (650) 736-1784

To the men and women of the Folsom control room at the California Independent System Operator and to reliability professionals everywhere who safeguard the critical systems upon which we all depend.

CONTENTS

ACKNOWLEDGMENTS

A LONG-TERM STUDY SUCH AS OURS INCURS considerable debt along the way. Most important, we thank Jim Detmers, Jim McIntosh, and Lonnie Rush of the California Independent System Operator (CAISO) for allowing us access to the control room floor, supporting our research over the years, and listening when we had something to say. Special thanks go to Dave Hawkins at CAISO for his encouragement at crucial stages of our research. Working with engineers Barney Speckman and Bob Stuart on an emergency response consultancy for the ISO added another layer to our appreciation of the complexity of the California grid. Our research assistants at CAISO, Patrick Getz, Hoai Dang, and Farakh Nasim, have been exemplary in developing, gathering, and analyzing the database used in Chapter 12, a database that also profited from the statistical help of Scott Fortmann-Roe. Specific thanks to CAISO's Steve Gillespie and all control room and operations personnel, including Joan Berglund and Kathleen Fernandez, who gave generously of their time and never lost patience with our stream of questions.

At the beginning of the research, we had the good fortune of working with researchers from Delft University of Technology in the Netherlands, particularly Michel van Eeten and Mark de Bruijne. Their input into our early thinking is evident here and in early articles we wrote together; they, however, share no blame for the direction in which we have taken that thinking. Their important work on reliability in telcoms, cybersecurity, and critical infrastructure interdependencies continues. We also owe a great deal to the high reliability organizations group at the University of California, Berkeley. Todd LaPorte took time to observe the CAISO control room and has served as a steady sounding board for our evolving ideas and analytic framework. An early lunch with Karlene Roberts was encouraging, as have been the comments of Gene Rochlin.

At other institutions, Peter Hirsch, Don Kondolean, and Joe Eto saw to the funding of a report on the initial research for the Electric Power Research Institute (EPRI) and the California Energy Commission, and they also made helpful comments on that report. Our book includes redrafted sections of the report as well as updated material from earlier articles:

- E. Roe, M. van Eeten, P. Schulman, and M. de Bruijne. 2002. *California's Electricity Restructuring: The Challenge of Providing Service and Grid Reliability.* Report prepared for the California Energy Commission, Lawrence Berkeley National Laboratory, and the Electric Power Research Institute. Palo Alto, CA: EPRI. We thank EPRI for permission to reprint and adapt material from the report, especially from its Chapter 9 (coauthored with Michel van Eeten), Appendix A (authored by Mark de Bruijne), and Appendix B.

- P. Schulman, E. Roe, M. van Eeten, and M. de Bruijne. 2004. "High Reliability and the Management of Critical Infrastructures." *Journal of Contingencies and Crisis Management* 12(1). We kindly acknowledge permission given by Blackwell Publishing to reproduce material.

- E. Roe, P. Schulman, M. van Eeten, and M. de Bruijne. 2005. "High Reliability Bandwidth Management in Large Technical Systems: Findings and Implications of Two Case Studies." *Journal of Public Administration Research and Theory* 15(2).

- P. Schulman and E. Roe. 2006. "Managing for Reliability in an Age of Terrorism." Chapter in Philip Auerswald, Lewis Branscomb, Todd M. LaPorte, and Erwann O. Michel-Kerjan (Eds.), *Seeds of Disaster, Roots of Response:*

How Private Action Can Reduce Public Vulnerability. New York: Cambridge University Press. We acknowledge permission to use material from Cambridge University Press.

- P. Schulman and E. Roe. 2007. "Dilemmas of Design and the Reliability of Critical Infrastructures." *Journal of Contingencies and Crisis Management* 15(1). We thank Blackwell Publishing for permission to reproduce this material.

We have given presentations to conferences on our findings, including the 5th International Conference on Technology, Policy and Innovation; the 24th Annual Research Conference of the Association for Public Policy Analysis and Management; a 2002 conference on the California electricity restructuring and crisis sponsored by the California Public Utilities Commission; the 2007 Conference on Critical Infrastructure Protection; and a 2005 Scandinavian Consortium for Organization Research (SCANCOR) Workshop on the structure and organization of government at Stanford University. Some of the very best feedback has been from the presentation of our interim findings to CAISO managers and through our many interviews.

We are genuinely grateful to all the people we interviewed over the years for this book. To maintain confidentiality, none of those interviewed is named in the text and two of our interviewees asked to remain entirely anonymous. We thank them as well as the others who also gave generously of their time to discuss topics with us through the years of research: Ziad Alaywan, Alan Amark, Massoud Amin, Ali Amirali, Steve Anners, Steve Auradou, Kevin Bakker, Diane J. Barney, Jamal Batakji, Barry Bauer, Larry Bellknap, Tracy Bibb, Chris Bossard, Boyd Busher, Ron Calvert, Russ Calvery, Sal Cardinale, Karen Cardwell, Rich Cashdollar, Sirajul Chowdhury, Jason Clay, Kevin Coffee, Robbie Cristillo, David Delparte, Terry P. Dennis, Jim Detmers, Carl Dobbs, Patrick Dorinson, Sanjay Dutta, Bill Ellard, Tami Elliot, Lisa Ellis, Joseph H. Eto, Paul Feeley, Mike Fling, Kellan Fluckiger, Mark Franks, Lee S. Friedman, Steve Gillespie, Mary Glas, Hector Gonzalez, Barbara Hale, Duane Haushild, Dave Hawkins, Christine Henry, Donna Hoop, Lauri Jones, Mike Karg, Stan K. Kataoka, Curtis Krankkala, Jamie J. Labban, Don LaDue, Eddie Ledesma, Stephen T. Lee, Yeaphanna LeMarr, Deborah LeVine, Lawrence Lingbloom, Deane Lyon, Stephen A. MacCool, Jeff MacDonald, Kim Malcolm, Jeffrey McDonald, Ed McGowan, Jim McIntosh, Jeffrey C. Miller, Marty Moran, Dale Murdock,

Randy Nunnink, Paul Olson, Shmuel S. Oren, Lori Parker, Dennis Peters, John Phipps, Sheri Pryor, Brian Rahman, Jason Regan, Guy Ridgely, Mitch Rodger, Mark Rothleder, Mike Starr, Steve St. Maire, Steven Stoft, Ed Street, Robert Stuart, Mark Stuckey, Robert Sullivan, Bob Therkelsen, Brian Thurston, Greg Tilliston, Nancy Traweek, Tim Van Blaricom, Greg Van Pelt, Chaise Vidal, Danielle Vockeroth Smith, Monique Weiland, Keith A. Wheeler, Mark Willis, Terry Winter, and Frank Wolak. This list does not include the nameless many we have talked to during CAISO-related workshops and training sessions over the years.

At the end of our work, Arjen Boin provided very detailed edits and comments on the draft manuscript, and for the support and suggestions of Margo Beth Crouppen at the Stanford University Press; David Horne, our copyeditor; and a second reviewer we are extremely grateful.

Conceptualizing and writing this book have been mutual and equal. As such, we together—and most certainly none of those listed above—are responsible for whatever errors of interpretation remain.

Finally, our respective families have borne the brunt of our work in terms of days away from home and hours at home processing this material. So to Louise and Scott and to Linda and Rebecca, we both say: Thank you for your loving patience and support.

May 2008   Emery Roe
                                                              Paul R. Schulman

CHAPTER I

THE STAKES

News Release (sometime in the future)
Reliability Crisis Leads to National Conference in the Capitol

The National Infrastructure Reliability Conference opened this morning in Washington, D.C. The conference was convened by the president in response to the growing number of high-profile failures among the nation's electricity grids, air traffic control operations, telecommunications systems, financial exchanges, and interstate water supplies. "We have to get to the bottom of why these critical systems are not meeting the reliability standards our citizens have a right to demand and expect," the president said in his opening remarks. "We have to reduce an unprecedented and unacceptable vulnerability."

The president's address also focused on the recent catastrophic failure of the western grid, which knocked out electricity for over thirty million households, disrupted ports and related transportation along the West Coast, and for more than a week interrupted major financial services and cell phone activity west of the Mississippi. The economic losses are estimated to be in the tens of billions.

The president addressed over eight hundred conference delegates. They included the CEOs and top executives of the largest corporations, which own over 85 percent of our nation's critical infrastructures. Also attending were leading engineers, economists, lawyers, and consultants who design these systems. Following the president's remarks, conference delegates listened to the keynote speech by the director of the National Academy of Engineering.

"We've invested billions of dollars, over the past two decades, designing and building the world's most advanced technical systems to provide essential services," she told delegates. "We've developed the most sophisticated risk assessment methods and designed the latest in safety systems. We've restructured our organizations to run these systems—streamlining them to operate with maximum efficiency to adapt to the changing requirements of new technology.

"Yet for all the investment," the director continued, "we have not realized proportionate improvements in the security and reliability of electricity, transportation, telecommunications, and financial services. Our national academy panel of experts now finds the nation vulnerable to more failures in our interdependent systems. The public is less confident today of infrastructure safety and dependability than they have ever been, and with good reason.

"We have lost something very important," the director concluded. "There were organizations in the last century that had far better records in running large, unforgiving technical systems. We need to recapture what they knew thirty years ago about running our complex infrastructures. The consequences are enormous if we don't."

A spirited debate followed.

"It's doubtful we've lost any real skills or information over the past quarter century," countered one of the country's leading electrical engineers. "Our expert systems and engineering models cover far more of the operation of complex technical systems than the design principles of the past or the 'gut' feelings of operators ever did." One Nobel laureate economist added, "Remember, many of the organizations of the past that ran these technical systems were rigid bureaucracies. They lacked the flexibility and incentives to adapt to changing infrastructure requirements."

The debate was especially heated in the afternoon conference panel, "Where to Now?" A prominent historian of science and technology gave a presentation on "high-reliability organizations." "Decades ago, a small group of researchers claimed to identify a set of organizations that made a special commitment to defy the odds and manage highly hazardous technical systems in air transport, nuclear power, and electric power with reliability as their highest objective. And these organizations did so with extremely high effectiveness," he argued. "Much of the research on these organizations was done in old-fashioned case studies quite difficult to locate now through our parallel cyberscans," added the historian. "But these studies, which focused more on organizational and management issues than on technology, could have considerable value in relation to present reliability problems."

"What is it they could possibly tell us?" questioned a well-known engineer from the floor. "What design principles can we distill from another era and a world so removed from our present technology and optimization methods?" The historian answered by describing the principal features of high reliability organizations.

"These organizations treated high reliability as a management challenge as much as a challenge in formal design. They didn't entirely trust the diagrams for their technical systems, nor the formal models and specifications describing how they should work. A great deal of attention was paid by members of these organizations to preparing for the unexpected—system conditions and events outside those specified in formal designs. Given this skepticism, a much larger role was played by operators and supporting personnel, who supplemented formal training with informal, experience-based perspectives on how these systems actually worked."

"This operator-and-discretion-based orientation would scarcely be possible today," interjected a panel member, "given our widespread self-referential expert control systems and flex-job practices throughout the economy. Worse yet, these older organizations clearly valued reliability of operations over optimization of performance. Their optimizing methodologies were primitive, and on many fronts these organizations didn't even try to optimize resource use or output, deeming it a potential threat to reliability."

An economist on the panel added that high reliability operations existed in conditions that artificially insulated them from competitive market pressures.

"Even those high reliability organizations exposed to market pressures were protected by regulatory frameworks that forced all of their competitors to make similar reliability investments and bear similar costs."

"We must recognize today," continued the panel economist, "that no modern, global supply-chain network can afford such excess capacity and inefficiency. Nor can we afford to return to those stultifying regulatory environments that depress technological innovation and destroy organizational flexibility."

"Quaint, and even noble, though these organizations may have been," another panel member summed up, "we can hardly expect engineers and economists to turn their backs on the contemporary world of formal design-based systems and reprogrammable organizations."

The conference continues tomorrow with the unveiling of the System Contingency Analyzer and Response Optimization Tool (SCAROT). This parallel-processing control system produces performance forecasts based on simultaneous analysis of hundreds of thousands of interdependent components across critical infrastructures. Every five seconds it issues intervention commands to better optimize intersystem performance. High-Reliability Products, a global LLC and one of the world's fastest growing multinationals, markets it.

THE PRECEDING NEWS ACCOUNT may seem fanciful, but we argue that something like it could very well occur, if we do not take more seriously the growing challenge of managing our critical systems. This book is about why such a conference has not been necessary—yet.

The interdependence of society's large technical infrastructures for electricity, telecommunications, financial services, and the like is well established (Dillon and Wright 2005; Conference on Critical Infrastructure Protection 2007). So too are demands for heightened performance of these complex networked systems, accompanied by technological and economic approaches to optimizing their operations. Increasingly common are technological strategies to protect against external threats and internal failures that might cascade through these systems (Pool 1997; Evan and Manion 2002).

Today is the high season of engineering and the historical hour of the economist in the world of critical infrastructures. This book lays out the case for an alternative focus—what we term "high reliability management."[1] We believe that many current approaches to design pose a danger to our critical services as significant as the prospect of natural disaster or human attack. In fact, some design and technological changes promising added security realize the opposite by hobbling management and resources that could better protect us. We demonstrate the importance of these managerial skills to our current and future safety and security. The skills of the people who manage our critical infrastructures are misunderstood and neglected, even by some in the organizations charged with the creation, operation, or regulation of critical infrastructures.

Our argument is founded on research conducted over many years to understand the challenges and competencies of high reliability organizations (HROs). To this we add ongoing research on electrical grid reliability. This research has been undertaken over a six-year period, including extensive interviews with managers, dispatchers, engineers, and regulators within private utilities and power generators, from organizations such as the California Independent System Operator (CAISO), the California Public Utilities Commission (CPUC), and the California Energy Commission (CEC). Our analysis and argument are also informed by countless hours of direct control room observation.[2]

RELIABILITY HAS BECOME a worldwide watchword of citizens, clients, leaders, and executives. For some, it means the constancy of service; for others, the safety of core activities and processes (LaPorte 1996). Increasingly, it means both *anticipation* and *resilience,* the ability of organizations to plan for shocks as well as to absorb and rebound from them in order to provide services safely and continuously. But putting these together in a single strategy is a formidable challenge.

The study of reliability in critical systems is the study of strategic balances that must be struck between efficiency and reliability, between learning by trial and error and the prevention of risky mistakes, and between maximizing our anticipation of shocks and maximizing our resilience to recover after them. *High Reliability Management* is about how these balances are achieved and sustained.

From our research into nuclear power plants, electricity grid control rooms, water system control rooms, and air traffic control centers, we sketch the specifics of the skill set and managerial approaches that promote reliability in settings where high reliability has long been the *sine qua non* not only of operational success but of organizational survival. We know that managers and executives, as well as system designers in varied settings, can learn useful lessons from the case material. Further, we show that in the absence of this understanding, many critical infrastructures are vulnerable to the very threats design and technology are meant to prevent. In demonstrating this, we tell the story of "reliability professionals"—a special group of professionals whose commitment and dedication make the difference on a daily basis between catastrophic failure of services we all depend on for life and livelihood and the routine functioning we have come to expect. Physicians call their life-threatening errors "never events," and reliability professionals are just as keen to avoid similar failures.

We direct our analysis to society's critical infrastructures because of their overwhelming importance. "Critical infrastructures" are core technical capabilities, along with the organizations that provide them, that enable the provision of a wide variety of social activities, goods, and services. Infrastructures in the domains of electricity, water resources, communications, transportation, and financial services are by their very nature multipurpose. They are critical in that they are necessary elements for more specific secondary systems and activities. If they fail, a wide variety of social and economic capacities are affected, with considerable economic consequences. The financial damage due to the August 2003 blackout in the northeastern United States alone is estimated to have been more than US$6 billion (de Bruijne 2006, 1).

Critical infrastructures have unusual properties that make them challenging to manage reliably. They are generally networked systems (de Bruijne 2006) with significant spatial dispersion, consisting of multiple organizations and varied interdependencies. Many of them, such as electric and transportation grids, have grown and developed by accretion, meaning they frequently consist of elements of varied age, design, and performance characteristics. The elements are difficult to characterize in a single model or consistent set of engineering or economic principles. To understand these systems and their reliability we must understand the special challenges they pose to management. In describing the challenges of high reliability management, we intend simultaneously to identify the practices, describe the professionalism, sketch out a new research field, and draw implications of our argument for more effective policy and management.

There are no shortcuts to high reliability, though the temptation to try to find them is strong. Our own concluding recommendations, if simply distilled into design principles or management recipes, will assuredly make these systems more brittle than current design strategies have already rendered them. Failing to understand and appreciate the practices, professionalism, and research findings associated with high reliability management means a continuation of what we believe are great risks associated with many policies, technical designs, and business fads current today.

OUR ANALYSIS is a cautionary tale cast around three propositions:

1. There is an important divergence between dominant approaches to designing systems and the process of managing them. The difference is primarily in cognitive orientation—ways of framing, knowing, and responding—among different groups of professionals.

   • In particular, a distinct set of reliability professionals can be found among operators and middle managers of complex technical systems—individuals who are more than usually competent and motivated to have things "turn out right" in the operation of complex systems. They balance learning by trial and error with the prevention of high-risk mistakes; they anticipate to avoid shocks but maintain an ability for quick response to them when they happen.

   • This balancing typically requires skills to work effectively in a special cognitive frame of reference between the general principles and deductive orientations of designers and the case-by-case, experience-based preoccupation of field operators. These concerns drive reliability professionals to manage in terms of the patterns they recognize at the systemwide level and the action scenarios they formulate for the local level.

2. High reliability management is focused less on safeguarding single-factor or invariant performance than on maintaining a set of key organizational processes through adjustments within upper and lower limits acceptable for management (what we call "bandwidths"). The boundaries of these

bandwidths can be reliably changed only in proportion to improvements in the special knowledge base of the system's reliability professionals.

3. Despite the vulnerabilities they generate, centralization and interdependency among the component parts of a complex technical system can actually be significant managerial resources for reliability. Notwithstanding recent preoccupations among designers to do away with these properties, they provide options with which reliability professionals can make key adjustments and preserve balances needed for resilience and anticipation in a complex technical system.

As we illustrate, each proposition represents a neglected perspective in current policy and technology approaches to the operations of large technical systems. In fact, high reliability management in many respects is about the management of errors associated with both technology and policy.

In California's electricity restructuring, for example, the stated theories of economists, regulators, and legislators that electricity markets would quickly evolve and attract new players in electricity generation as well as keep wholesale prices down by means of market forces did not prove correct. In reality, managers of the grid confronted quite the opposite. They faced an unstable cycle of design errors leading to underperformance leading to ever-more frantic efforts at redesign.[3]

The architects of electricity restructuring were quick to say that their design was not really tried. For example, the retail market for electricity was not deregulated the way the wholesale market was. But smart and talented economists, policymakers, and regulators over a sustained period of time gave deregulated energy markets their best shot in California, and there were still major unforeseen and undesirable consequences they could neither predict nor control. Very clever reformers failed to forecast the first-order, let alone second-order, consequences of their policies, as we document in the book. Here, as in other cases, engineers, regulators, and economists were sorely undereducated when it came to the management requirements for highly reliable performance. We hope through this book to fill that education gap by making it clear just what the management requirements for high reliability performance entail.

We argue that no strategy of policy, technology, or markets can ensure reliability on its own without a strong management base. The importance of that base is seen everywhere in the California example. Throughout the electricity crisis and the long aftermath induced by restructuring, the lights by and large have stayed on in the state. Why? Quite simply, the California Independent System Operator (CAISO) and the distribution utilities, including Pacific Gas & Electric (PG&E) and Southern California Edison (SCE), have taken reliability seriously, when others did not. How so?

First, according to economists, reliability is only one attribute of electricity and can be traded off against other attributes, such as how cheap or environmentally "clean" the energy is. But that clearly has not been the case. Reliable electricity (or water or telecommunications or so on) *defines* the critical infrastructure, and we as a society are unwilling to trade off reliability against the service's other features.

Second, advocates of major redesigns in our critical infrastructures have argued that reform, through either new technologies, markets, or policies, would significantly reduce the organizational burden of coordinating complex interdependencies. Not so. Deregulated energy markets, to name but one case, have created a far greater task of coordinating interdependency among market participants. This burden has fallen upon key focal organizations such as CAISO.

Finally, the real experiment in this infrastructure reform has not been the policy or the new technology itself, but something altogether more disturbing: a scrambling and reshuffling of institutions—single integrated utilities on the one hand and entirely new organizations on the other—all on the premise that organizations can be created or dismantled at will, without undermining service reliability in the process. The broad implications of this monumental conceit are exposed in this book.

It is said of Americans that they hate "regulation" but at the same time demand all manner of safeguards to ensure their own personal safety, health, and well-being. Our critical infrastructures provide a clear example of this demand. Americans, not just Californians, have shown themselves willing and able to spend billions upon billions of dollars in the name of ensuring the reliability of critical infrastructures—think of Y2K. Such transfers of income demonstrate that critical service reliability is not like any other "attribute" we know. High reliability is not just one more quality feature in the hedonic price for electricity. In fact, it is a foundation for the operation of society, not simply an attribute of a service, in the same way a society's distribution of assets is a foundation from which any set of prices are derived. Substantially change asset distribution, and you change the prices; substantially change service reliability of our critical infrastructures, and you change their character as assets. Critical infrastructures have become so intertwined, and we, as a society, are so dependent on their always-on reliability, that high reliability has become a background condition, much like the framework of contract law, that *enables* market transactions.

As we show, high reliability in real time is not a bargainable commodity, nor is it sensibly traded-off by individual consumers in order to cheapen the costs of service. In real time when it matters the most, reliability is not exchanged or substituted for something else; otherwise those services would not be critical. This lesson was learned dramatically and at great cost in the California electricity crisis. The state budget surplus disappeared as the governor and his administration spent it on buying high-priced "spot-market" electricity, because not enough prescheduled power was being bid into the new energy markets to keep the lights on. As this was unfolding, a sign—"Reliability Through Markets"—was unceremoniously removed from the CAISO control room. It actually should have read "Markets Through Reliability." The evaporation of the California budget surplus (over $12 billion) and the recall of the state's governor testify to the foundational role of infrastructure reliability in modern society.

WHEN WE LEAVE CALIFORNIA, the challenges to critical infrastructures, from within and outside, remain the same. It is essential that all of us understand why.

Imagine a coordinated attack by terrorists striking at major electric power transmission lines and facilities in strategic places throughout the American Midwest and Northeast. They are able to knock out nearly 70,000 megawatts of peak-load electrical capacity and throw more than fifty million people into darkness over a 240,000 kilometer area in the United States and Canada. Without electric power a variety of other critical services fail, including water supplies and hospital facilities, as well as major financial markets over the globe. Ultimately, security systems become disabled, leaving key infrastructures vulnerable to additional terrorist attacks.

By this point, you may have already guessed that many of these conditions actually existed during the Northeast blackouts of August 14, 2003. The outages were caused not by terrorists but by the failure of electric transmission systems themselves, without hostile intervention. Although power was restored quickly in some areas, other portions of major metropolitan regions were without power for over twenty-four hours, and some areas had service interruptions for several days. It could have been worse. An earlier report issued in 2002 by a task force headed by former senators Gary Hart and Warren Rudman concluded that as a consequence of a coordinated terrorist attack, because of the lack of replacement parts for aged or customized equipment, "acute [power] shortages could mandate rolling blackouts for as long as several years" (Regalado and Fields 2003, A3). On November 4, 2006, the shutting down of a high-voltage line over a river in Germany to allow a ship to pass led to a chain-reaction set of outages that plunged ten million people in six European countries into darkness.

It is not only electric grids we have to worry about. We confront information networks under assault by computer viruses and hackers, large-scale transportation systems and water supplies open to terrorist attack—even the prospect of electronic voting exposes us to all manner of fraud and undetected error. Surprisingly, it is not expanding the reach of these complex systems but rather safeguarding their reliability that has become a great preoccupation of the twenty-first century.

At the same time, this preoccupation is often misguided in ways that are not fully appreciated. System designers and policymakers assume that the key to reliability lies in hardening our infrastructures so that they better resist outside attack. It is said that if we make these large technical systems more failsafe and foolproof or less tightly coupled or even more physically dispersed, we will improve their reliability (National Research Council 2002; Farrell, Lave, and Morgan 2002; Perrow 1999 [1984]). As one engineer has contended, "I try to design systems that are not only foolproof but damned foolproof, so even a damned fool can't screw them up."

It seems every week we hear of new data processing, electronic communications, and security systems that are planned or have already been put into place to increase reliability in the face of terrorist threats. More design changes are planned for electricity grids and air traffic control systems (for example, Apt, Lave, Talukdar, Morgan, and Ilic 2004). Recent public discussions of business continuity have focused almost exclusively on design solutions to problems of businesses' protection or recovery from external threat (for example, *Financial Times* 2005). From our perspective, wrapping a patient suffering from internal bleeding in body armor of the latest style is not therapeutic.

We intend to show that the key to increased reliability for our electricity grids, water supplies, telcoms, transportation systems, and financial services, among others, lies not in the pursuit of optimal designs and fail-safe technologies for large complex systems but, rather, in their careful management. Unfortunately, there is a paradox at the core of this management. The very skills of high reliability management described in this book mask the vulnerability of that management to challenges and stresses, including those induced by misdirected designs and policy interventions. The professional attitude of key personnel and their virtuosity at working around design errors and rescuing situations in real time means that few signals are sent by these managers that conditions are worsening, until major failure actually occurs. Even higher management in their own organizations, as we will demonstrate, may not see how close to the edge the system is operating with respect to maintaining reliability.

Our intent is to signal what is happening in and to our critical infrastructures today. Rather than wait for major failures to communicate the risks, we offer a detailed look at the world of high reliability management, with careful case descriptions and close analysis of the skills at work. In the process, we offer a new method for measuring precursor conditions that serve as early-warning indicators of approaching edges to capacities for high reliability management.

THIS BOOK IS ORGANIZED into three sections. The principal focus of Part One is on the change over time in the reliability management of the California Independent System Operator (CAISO) as the transmission manager of California's high-voltage electrical grid, one of the world's most important electricity systems. To our knowledge, no other critical infrastructure control room has been examined as intensively with respect to managing for high reliability and over such a lengthy period of time (2001–2008).

We also discuss in Part One theories relevant to the challenge of high reliability, including "normal accident" theory and the theory of high reliability organizations (HROs). We show the limitations of these theories in accounting for what we have observed, and we present our own framework for understanding high reliability management. This framework highlights the crucial role of reliability professionals—control operators, key technical department heads, and support personnel—whose special cognitive skills and flexible performance modes maintain reliable operations even in the face of widely varying and unpredictable conditions.

Part Two puts our analysis of key concepts, practices, and issues in high reliability management into a strategic perspective. Topics covered are (1) the critical balance between trial-and-error learning and failure-free performance in large technical systems; (2) strategies of managing performance fluctuations within controlled upper and lower limits and margins (bandwidths) as opposed to strategies for invariant performance; (3) the special domain of operational risks as opposed to analyzed risk, a domain in which risk *seeking* can enhance reliability; (4) the cognitive meaning of anticipation, resilience, robustness, and recovery and their operational trade-offs; and (5) the special threats to high reliability management posed by current approaches to technical design. Part Two ends with a chapter on indicators for key concepts, including the identification, measurement, and assessment of performance edges in high reliability settings.

Part Three moves the analysis to the wider context of critical infrastructures and the implications of our findings for the high reliability management of infrastructures in other social and organizational settings. National defense and homeland security are given special attention. We conclude with an examination of ways to support and protect reliability professionals, so as to ensure the provision of critical services in the future. It is a great irony that, while many economists are calling for greater efficiencies in critical service provision, and many engineers for greater capacity and effectiveness in the design of these services, the most underutilized resource we have as a society is the professionals who run these systems. They actively and consistently protect our infrastructures against disturbances, failures, and mistakes that could bring them down, including errors at policy and supervisory levels. They work, often heroically, against odds that many in the public, academia, and government can hardly appreciate.

PART I

THE CASE STUDY
California Independent System Operator, 2001–2006

Iн JANUARY 2005, WE MET WITH THE VICE president for operations and department heads of the California Independent System Operator (CAISO), an organization charged with the management of most of the California high-voltage electricity grid with its estimated sixty-five million consumers. The meeting was called as a result of a report we had written to the VP describing what we were seeing recently within his organization. We had interrupted our ongoing CAISO research, which had begun in 2001, to write that report.

What we wrote alarmed him. We had concluded that CAISO could no longer ensure the high reliability of the grid and thus electrical service in California. We wrote the report because we were very concerned by what we were observing in CAISO's control room and its implications given our understanding of high reliability management. The VP acted on our report immediately. He initiated the January meeting and follow-up discussions in February 2005 to review our major points and suggestions. Along with other findings, these discussions among CAISO's other executive officers, department heads, control room managers, and operators led to some of the changes we recommended.

This series of events is highly unusual in organization research. It is rare for researchers in the social sciences to intervene in organizations they are studying. But it became clear to us that the significant problems we were observing in CAISO's management could be revealed and understood by the very analytic framework CAISO operators and managers had helped us devise in our first phase of research beginning in 2001. It was this analysis that signaled hazards ahead and what could be done to address them.

The best way to tell the full story is to flash back to CAISO in 2001. It was the height of the California energy crisis, when we began our firsthand, on-site research and observations. The story we begin there is about the skills required to maintain critical services safely and continuously in the face of design errors and illusions at the level of public policy. It is a story about keeping the lights on while coming closer and closer to the edge of resources and capacities to do so. Moreover, the story parallels what is happening in other key infrastructures today, as we will see.

On-site research with our colleagues began in April 2001, when CAISO was a fairly new organization managing the grid in the midst of California's bold and controversial restructuring of the electricity sector. This "deregulation" (more formally, restructuring) began in 1998 with the passage of a sweeping measure in the California legislature, AB 1890. This was a massive policy departure from a system of large, regulated-monopoly, investor-owned utilities that controlled power generation, transmission, and distribution, as well as sales and billing. It was now to be a system of wholesale price deregulation, creating wholesale electricity markets with multiple competitive suppliers. The restructuring also mandated the divestment by the utilities of much of their generating capacity (except nuclear and hydropower) and the transferal of grid management responsibility from the utilities to a new grid management organization, the California Independent System Operator.

Our research over the coming years came to embrace three distinct developmental periods in the history of California's electricity system. The first was one of environmental turbulence and rapid organizational growth for CAISO. This period encompassed the state's electricity crisis of 2000–2001, during which power shortages developed as wholesale markets failed to produce needed generation. During this period one major utility filed for bankruptcy and another teetered on bankruptcy. A "power exchange" created to manage the emerging power market went bankrupt, wholesale prices skyrocketed, and the state committed $12.5 billion (and nearly its entire budgetary surplus) to purchase power in the very expensive spot market in order to keep the lights on (Sweeney 2002). The state did not let failed markets undermine reliable electricity and the economy and society that depended on it.

The second period featured the efforts of external players to stabilize the environment of CAISO. This includes the efforts of the then governor of California to ensure electricity supply and price through the signing of long-term power contracts with generators in Mexico and other out-of-state locations. It also includes new regulatory policies by the Federal Energy Regulatory Commission (FERC) to enlarge regulatory mandates, which CAISO could invoke to require generators to provide power (on a "must-offer" basis) when needed for grid reliability.

The third period of development is one in which CAISO itself has undertaken initiatives to stabilize and expand its institutional niche in the restructured electricity sector. Senior grid managers had no intention of continuing to be at the mercy of conditions proven to be inimical to the cost-effective management of grid and service reliability. Major programs were initiated to redesign and gain greater control over electricity markets in the state. In addition, new software has been introduced to monitor the state of the grid and to automate grid transactions at faster speeds with multiparameter computations in order to optimize dispatch decisions with respect to power availability and prices. A new CEO has been brought on, and the organization was downsized in a 2006 realignment.

As the reader will see, two constants remained throughout these developments: the incompleteness of formal designs imposed upon operations relative to the wide variety of grid conditions actually confronted by operators, and the inability of CAISO to transform itself fully from an organization into an institution with stable political support and acceptance (Selznick 1957). The impact of these factors can be seen everywhere.

In addition to the bankruptcies already mentioned, other high-profile events witnessed during our case study period (2001–2006) included the recall of the state's governor, resulting at least partly from dissatisfaction over the energy crisis and the cost of the long-term energy contracts he signed; the documented use of market power by major power companies such as Enron to increase prices; and the bankruptcy of Enron and the convictions that followed from manipulated energy markets. There have also been unrelenting changes in the technology of the grid—new transmission lines, automated relays, and emergency switching systems—along with organizational changes within CAISO as it has been continually challenged in developing and institutionalizing its state, western, and national niche in the field of restructured electricity transmission.

During the entire case study period and beyond we have been given extraordinary access, as researchers, to CAISO control room operations on a daily basis, to frequent meetings at many organizational levels, and to many key individuals for interviews (many over multiple periods). At no point were we asked to limit our study or modify our findings. At no point were we asked to leave a meeting or not listen to a conversation. It is unlikely that any large organization with such high reliability mandates has been as intensively and extensively studied.

CHAPTER 2

CAISO IN 2001
The Electricity Crisis from Inside the Control Room[1]

THE CALIFORNIA INDEPENDENT SYSTEM Operator began life as an organization mandated to provide highly reliable electricity transmission but working under the error-filled design and regulatory confusion of California's electricity restructuring. This chapter shows the amazing ability control room operators had to rescue electricity flows hour by hour during the 2000–2001 electricity crisis. We identify the special skills demonstrated by operators in the control room and support staff for ensuring reliability in real time as opposed to what many engineers, lawyers, and economists had anticipated for the electricity system in its idealized restructured form.

FORTUNE MAGAZINE RANKED the California electricity crisis and Enron's collapse as 2001's major events right after 9/11 and economic recession, concluding that the crisis "left the deregulation movement in tatters" (Wheat

2001). Much of the California energy debacle is well known: the theoretical promise of deregulation (see for example, California Public Utilities Commission 1993) versus the reality of the crisis (for example, California State Auditor 2001; Congressional Budget Office 2001). That crisis has become an exemplar of "policy fiascoes" and "implementation gaps" (Bovens and 't Hart 1996; Hill and Hupe 2002).[2]

Our summary of events can be brief. The state implemented in 1998 a major restructuring in its electricity generation, transmission, and distribution. California moved from a set of large integrated utilities that owned and operated the generation facilities, the transmission lines, and the distribution and billing systems, and that set retail prices under a cost-based regulatory system, to a market-based system consisting of independent generators who sell their power on wholesale markets to distributors, who then sell it (in a regulated market) to retail customers.

The distribution utilities were compelled to sell off most of their generating capacity (except for nuclear and hydropower sources) and to place their transmission lines under the control of a new organization, the California Independent System Operator (CAISO), which assumed responsibility for managing a new statewide high-voltage electrical grid. California was the first state in the United States to create a separate independent system operator to control the transmission facilities owned by major distribution utilities (Kahn and Lynch 2000, 7).

The grid managed by CAISO represents primarily the merger of separate grids of the state's three major utilities, Pacific Gas & Electric (PG&E) in the north and Southern California Edison (SCE) and San Diego Gas & Electric (SDG&E) in the south. In 1996, these investor-owned utilities accounted for 75 percent of all customers and all state power sales (Weare 2003, 7). The head office and main control room of CAISO were located in the city of Folsom, just northeast of Sacramento.

THE CALIFORNIA ISO was built "from the ground up" within a short time frame. Established by legislation in 1996, it began operations in 1998. The delay was caused by problems in the startup (detailed in De Bruijne 2006). The markets and organizations were developed and created from scratch in little more than nine months, and many problems were yet to be solved at the time operations began (Joskow 2000b, 18; Blumstein, Friedman, and Green 2002, 25).

CAISO hired some two hundred employees and built its organization and equipment "[c]ompletely independent from any market participant—including the major California utilities," nevertheless requiring "full functionality at day one" (Alaywan 2000, 71). As an organization without precursors or precedent in California, CAISO "had to procure a building, hire an organization, and create an infrastructure for a complex leading-edge market. They had to design their protocols simultaneously with developing their systems" (The Structure Group 2004, 8).

Restructuring meant that California's electricity sector was to be composed of a novel mixture of private and public organizations that were involved in the provision of electricity. Electricity generation would become the business of competitive, independent electricity generators. These private generators were now to consist primarily of the (fossil fuel) generating plants that had previously belonged to the utilities. To this end, a new intermediary group was to enter the electricity industry through the private sector: the electricity schedulers, also known as energy brokers or, more commonly, scheduling coordinators (SCs). The task of SCs was to match electricity supply and demand in the new market setting. Matched blocks of supply and demand were to be submitted to the grid operator, CAISO, which calculated whether or not the schedules met the capabilities of the high-voltage transmission grid.

CAISO, to which operational control of the utility-owned high-voltage transmission grid had been transferred, was to be a hybrid, not-for-profit, public-benefit corporation. It mission was clear: "The ISO gained control of the statewide grid and is responsible for the reliable operation of the system that supplies wholesale power to the State's investor-owned utilities" (California State Auditor 2001, 7). CAISO's primary responsibility is to balance load and generation throughout the grid so as to guarantee the reliable, high-quality provision of electricity. "Load" is the end use (commonly, the demand) for electricity, and "generation" is the electricity to meet that load, both of which must be balanced (that is, equal to each other) within prescribed periods of time, or otherwise service delivery is interrupted as the grid physically fails or collapses.[3] To do this, CAISO can make necessary adjustments in the amount of electricity fed into the grid by ordering electricity generators to increase or decrease their output.

The restructuring envisioned CAISO making these adjustments through the use of a real-time "imbalance" market, a "congestion" market, and an "ancillary services" market. The real-time imbalance market was to enable CAISO to increase or decrease ("inc" and "dec") generation in real time in order to balance load and generation. As originally designed, the congestion market was to ensure the efficient and optimal use of the high-voltage electricity grid by comparing in advance what had been scheduled for transmission with the physical capabilities of the CAISO transmission grid. The ancillary services market consisted primarily of the following types of electricity generation:

- *Regulation:* generation that was already up and running and could be increased or decreased instantly to keep supply and demand in balance

- *Spinning reserves:* generation that was running, with additional capacity that could be dispatched within minutes

- *Nonspinning reserves:* generation that was not running, but which could be brought online within ten minutes

- *Replacement reserves:* generation that could begin contributing to the grid within an hour.

The ancillary services market was originally designed to accommodate about 3 percent to 5 percent of the total electricity needs of the state and provide CAISO with the necessary reserves to reliably operate the high-voltage electricity grid. When CAISO needed more energy than what was bid in the ancillary services and real-time imbalance markets, it could make "out of market" (OOM) purchases, typically from generators in other states or municipal utilities.

To ensure that CAISO would be provided with sufficient capabilities to balance load and generation, a system of warnings and emergency alerts was devised. If forecasted reserves for the next day were projected to fall below 7 percent of the projected load, CAISO would issue an Alert, and generators would be asked to increase their power bids into the market. If forecasted reserves during the day fell below 7 percent, CAISO would issue a Warning, and

CAISO would start buying these supplies directly. When actual reserves fell below 7 percent, 5 percent, and 1.5 percent respectively, CAISO would issue a Stage 1 emergency declaration (public appeals and other measures to increase supply and decrease demand), Stage 2 emergency ("interruptible-load" customers curtailed), and finally a Stage 3 emergency, under which distribution utilities would be instructed to black out even firm-load customers to reduce load to keep the system from crashing (Kahn and Lynch 2000, 6). The three markets, plus the emergency declaration system, were to enable CAISO to operate the electricity system—continuously balancing load and generation and providing the necessary resources to meet potential threats to service and grid reliability.

Markets were to be the main coordinating mechanisms for grid operations. In theory, market transactions would result in the electricity schedules that would be the basis of grid operations. A sign with the phrase "Reliability Through Markets" was posted prominently on a wall of the CAISO control room. That premise proved to be entirely unfounded. Instead, in the world of real-time market transactions, large energy suppliers ended up exercising market power, backed up by the threat of service interruptions, and set prices without benefit of market competition.

The defining feature of the California electricity crisis was a substantial increase in real-time, spot-market purchases of electricity at the last minute (Weare 2003; Duane 2002; Lee 2001). Operators told us about days that started with major portions of the forecasted load still not prescheduled and with the predictability of operations significantly diminished. One lead official in CAISO market operations explained, "The biggest problem . . . was that we had days where load is forecasted to be 42,000 megawatts [MW], but our scheduled resources in the morning were 32,000MW, leaving us 10,000MW short that day. How do we deal with this? Can we put a unit online? No we can't. Can we purchase energy prior to real time? No we can't. . . . Ninety-nine percent of the planning has to be done prior to real time. Real time is only time to react to what you missed. Real time is not 'I'm short 10,000MW in the day ahead and I'm not doing anything.' Most of the time things did come together, but at a very high price."[4] By the end of the California electricity crisis, the markets as originally designed had atrophied or disappeared. With the loss of nearly the entire state budgetary surplus, the electricity restructuring may well have done more financial damage to the California state economy than any single terrorist attack on the grid could ever do.[5]

THE POPULAR VIEW of California's electricity crisis is that its defining feature was the *unreliability* of electricity provision. Yet for all the design-induced errors and expense, during the electricity crisis the lights by and large stayed on against enormous odds few in the public understand. The real story of California's restructuring is that the lights did stay on, and they stayed on through many close calls because of the dedication of, cooperation among, and virtuosity of controllers, supervisors, and operators in California's restructured electricity network in the middle of the "perfect storm in electricity," as many termed it.

Notwithstanding the popular view of rolling blackouts sweeping California during its electricity crisis, in aggregate terms—both in hours and megawatts—blackouts were minimal. Rolling blackouts occurred on seven days during 2001, accounting for no more than twenty-seven hours. Load shedding ranged from 300 to 1,000MW in a state whose total daily load averaged in the upper 20,000 to lower 30,000MW range. In total, the aggregate amount of megawatts actually shed during the rolling blackouts amounted to slightly more than 11,500 megawatt-hours (MWh). The California electricity crisis had the effect of less than an hour's worth of outage for every household in the state in 2001 (de Bruijne 2006, 143).

Understanding why the lights stayed on in the crisis reveals the importance of high reliability management—and not just in electricity, but also far beyond. As has often been recorded, electricity underlies the high reliability of other large technical systems, including telecommunications, water, and financial services, also considered critical infrastructures.

Understanding how the lights stayed on underscores the skills of those who keep our critical services reliable, and teaches us—the public and our policymakers—what actually improves and enhances their skills as well as degrades them, no matter how well-intended the "improvements and enhancements" may be. Over the years of research and stakeholder interviews, we have found few who appreciate the necessity of having a cadre of skilled operators, the class of people we call reliability professionals, who can meet our high reliability mandates.

AS WE AND OUR COLLEAGUES began work at the California Independent System Operator, our original research questions became clear. How does an organization maintain reliable provision of electricity in a networked setting, in which both those owning the generation and those handling distribution of electricity (the privatized generators and reorganized "public" utilities) have competing goals and interests and there is an absence of ongoing command and control to ensure reliable electricity services? How can a network of deregulated generators, transmission managers, and distribution utilities achieve highly reliable electricity supplies over time, with high reliability defined as the safe and continuous provision of electricity even during peak demand times? How can a focal organization in that network, such as the grid transmission manager CAISO, ensure high reliability grid operation in real time, when load and generation must be balanced in the current or next hour rather than in the day-ahead or hour-ahead markets created through restructuring?

With those research questions in mind, we discussed with CAISO officials our entering the organization and interviewing key staff as well as undertaking direct control room observations. Eventually, we obtained approval and began our interviews. Little did we know that by the time we first entered the CAISO control room in April 2001, we would be witnessing the California electricity crisis, firsthand and from a perspective no others had.

HERE IS WHAT WE SAW when we first walked into CAISO's control room: a semi-circular, high-ceilinged room with a large wall-to-wall, ceiling-to-floor mosaic map-board giving a schematic overview of California's high-voltage electricity grid. That curving, full-length panel was impressive. Figure 2.1 shows the most recent CAISO control room.

After all our visits, the map-board still grabs our attention and that of almost every visitor. The actual configuration of the room has changed little during our six-year research period.

The map-board is the front wall of the control room. To its left side (when one faces the front wall) is a large, multifunctional video monitor, which displays the status of the most important indicators and parameters of California's

FIGURE 2.1. CAISO control room

SOURCE: By courtesy of the California ISO.

electricity grid renewed every few seconds by software. Control of the grid centers around four critical variables: frequency, Area Control Error, voltage, and electric power flow across the grid in megawatts. In relation to these control variables the key parameters for high reliability are the following:

1. *Voltage support.* It is important to maintain key voltages along the entire generation, transmission, and distribution system. Too low a voltage over time could cause the transmission grid to collapse. Too high a voltage can destroy equipment on the user side of the system.

2. *Frequency stabilization.* The established frequency standard for electric power (60.0 Hertz in the United States) must also be maintained across the entire high-voltage electrical grid. Too low a frequency can cause damage to user equipment as well as the loss of accuracy of clocks and other precision electrical devices that are designed to operate at that frequency. Small fluctuations must be corrected by minute frequency changes in the other direction in order to maintain the accuracy of clocks. Above the control room's left-side video monitor, a small panel shows the frequency in California's electricity grid. The WSCC (Western States Coordinating Council in 2001; since renamed WECC, the Western Electricity Coordinating Council) set standards covering frequency (Control Performance Standard 1, or CPS1) that it applies to CAISO.

3. *Load and generation balance.* The most critical reliability requirement is a close balance between load and generation across the grid. More load than generation can cause the grid to collapse, leading to widespread outages, as have occurred periodically in regions across the United States. Too much generation relative to load can burn out parts of the grid and cause the loss of billions of dollars of physical assets. The balance is monitored through the Area Control Error (ACE), which is also observed on the left-side video monitor. The ACE must fall within WECC-established limits for each ten-minute interval during every hour (Control Performance Standard 2, or CPS2).

4. *Path protection.* Another critical reliability challenge is the protection of critical transmission pathways along the grid. Load limits must be observed, and their temperature must be monitored and overheating avoided. Otherwise the lines will sag and potentially come into contact with the

ground or objects that could cause fires or fatal accidents. NERC, then the North American Electric Reliability Council (now Corporation), has established load and thermal limits for path protection.

Running along the right side of the control room is a large glass wall, which looks into the control center's main meeting room. That room, the "fishbowl," is equipped with a range of equipment enabling it to be used for emergency conference calls and meetings of the support staff during Stage 1, 2, or 3 emergencies.

In 2001, the control room proper had thirteen computer consoles for its operators. Each console had a series of monitors to provide the control room operator with the software to perform his or her function. During hurried periods of the day, operators were often on the phone when not working at their monitors. A great deal of their work was face-to-screen, even during phone calls. Figure 2.2 shows a schematic of the control room as it looked then.

Much of what the operators at the control room consoles do at the time of writing (2007) is very different from what we saw in 2001. The changes, many of which flowed from the California electricity crisis, are described in later chapters. Here it is important to understand what the baseline of operations and observations are for this book.

What we have numbered console 1 was occupied by the then WSCC "security coordinator" (now known as a reliability coordinator) for the California power area. This individual was (and up to the time of writing still is) responsible for monitoring this part of the western grid. The western grid

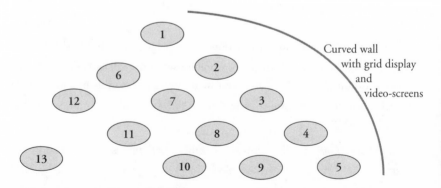

FIGURE 2.2. Schematic display of CAISO's Folsom control room at the time of initial research

includes not only California but also the western United States and parts of Canada's and Mexico's high-voltage electricity grid. This portion of the U.S. grid is "stability limited," which means that the system has inherent properties that make it unstable after certain limits (for example, with respect to maximum load on transmission lines) are exceeded. To prevent local or regional outages from cascading and bringing down the western grid, the system is equipped with Remedial Action Schemes (RASs) that automatically trip and "island" the grid into smaller subsystems, if and when critical problems have been detected on important parts of the system.

To further ensure the reliable operation of the western electricity grid, the WECC subscribes to Minimum Operating Reliability Criteria (MORC). MORC apply to all system operators under all conditions. For example, one of the WECC-MORC standards sets the system operator's operating reserves at the aforementioned minimum of 7 percent of its total load. The security coordinator ensures that the system operators in the region meet the various voluntary NERC and WECC reliability standards and operate the grid within the allowed performance bandwidths. Violations are recorded and subject to monetary penalties by the WECC. The security coordinator also acts as liaison and coordinator between CAISO and neighboring control areas to respond to large-scale problems that potentially threaten the stability of the grids of multiple system operators.

The transmission dispatchers, who monitored and managed the status of CAISO's high-voltage transmission grid, occupied consoles 2 and 3. Changes in transmission directly affect the load constraints and the performance characteristics of the rest of grid. Transmission dispatchers processed and coordinated responses to any threats to the transmission system, such as "overloading" of transmission lines along with planned or unplanned transmission outages. All changes in the status of (major components of) the grid, such as outages of power plants or failures of transmission lines, were processed and logged in the Scheduling Logging for the ISO of California (SLIC) system by staff. The SLIC system is a repository of a wide range of information, which we use in our statistical analysis later in this book.

At the time of our 2001 observations, console 4 was occupied by CERS (California Energy Resources Scheduling) personnel, who purchased electricity on behalf of the state for distribution by the utilities during the electricity crisis. CERS, actually a unit administered by the State's Department of Water Resources, soon established itself as one of the largest electricity buyers in the nation, purchasing about 90 percent of the distribution utilities' wholesale electricity needs, which amounted to about one third of California's total power use (Congressional Budget Office 2001, 28). From January to May 2001 alone, the State of California paid an estimated $8 billion to keep California's lights on, severely tapping into its financial reserves (Joskow 2001a). CERS later began to negotiate large forward electricity contracts against set rates, committing the state to long-term electricity contracts. The contracts were said to amount to between $40 billion and $50 billion, and after market rates fell, pressure to renegotiate these contracts mounted. CERS has long since left the control room, and its contracts were one of the many reasons for then Governor Gray Davis's forced exit from the state's active political scene.

Console 5 was used as an emergency support, backup, and training desk. Consoles 6 and 7 represented the day-ahead and hour-ahead market desks, respectively. The markets were to enable CAISO to operate the electricity system by continuously balancing load and generation and by providing the resources to meet any potential threat to service and grid reliability. The market operations were thus to be closely interwoven with CAISO's system control operations. It bears repeating that market transactions were to guide and coordinate grid management. What this meant for the control room was that the congestion market and the ancillary services market were both run in a day-ahead market and an hour-ahead market. The energy imbalance market was run in real time.

The day-ahead grid resource coordinator (at console 6) collected the schedules for balanced load and generation filed by the SCs and used a congestion-management program to determine if the schedules could be met given the physical constraints of the high-voltage electricity grid. If the grid could not accommodate the schedules, transmission lines would be congested. CAISO tried to facilitate both control of the network and the more efficient allocation of electricity by accepting "adjustment bids" to sell or buy more power to reduce congestion or by instituting charges if market participants continued to schedule power over transmission lines that were heavily used.

A considerable amount of strategic behavior and gaming by SCs occurred in the congestion market during the California electricity crisis. (Enron was one of the state's notorious SCs.) In brief, the electricity schedules were aggregated, processed, compared with the electricity grid constraints, adjusted, and redistributed to the market participants as the Final Day Ahead Schedule. The day-ahead market was meant to close at 10 A.M. each morning for the day ahead, though it has been left open later in order to allow late schedules.

Next to the day-ahead desk was the grid resource coordinator for the hour-ahead market (console 7). This market was intended to allow market parties and CAISO to adjust the amount of electricity they have purchased in both the ancillary services and congestion markets in light of any changes since the day-ahead market was closed. In reality, the hour-ahead market was a misnomer, as it procured electricity three hours before actually dispatching the energy. The extra time was needed to allow CAISO's computers to perform the highly complex task of (re)computing, (re)calculating, and adjusting the electricity schedules and congestion-management charges.

Consoles 8 and 9 were staffed by CAISO's real-time schedulers. They were responsible for monitoring, coordinating, and managing the real-time electricity flows on California's high-voltage transmission interconnections with neighboring electricity systems (interties). The loads and changes on these ties (and thus loads and flows in other control areas and their grids) directly affect the stability of California's electricity grid and need to be matched with California's internal electricity flow patterns to avoid congestion and system disturbances. At the time of our research, twenty-five such interties connected the California grid with adjacent grids, the most important being the California-Oregon Intertie (COI), which connects California to large hydropower resources in the northwest. A great deal of the state's electricity needs were, and still are, imported over the interties.

Apart from the constraints that the interties and other control areas pose, grid operations in California are influenced by the inability to transport large electricity loads between different parts of the state. A lack of transmission capacity inside California has taxed the ability of CAISO to maintain the reliability of California's electricity grid. By far the most important bottleneck in California's electricity network at the time of our 2001 research was the set of transmission lines designated as "Path 15." Path 15 represented the only way to transport large volumes of electricity between northern and southern California and was wholly inadequate to perform that task.

Problems were compounded by the relatively uneven distribution of both the type and number of power plants and the distribution of load between the north and the south, which made the path critically vulnerable. During periods of congestion and severe strains on the grid, Path 15 was not able to transfer the amounts of electricity needed between these two areas. When that situation occurred, California's electricity grid had to be managed as if it were two independent grids, called North of Path 15 (NP15) and South of Path 15 (SP15). This meant that electricity prices between the different zones differed substantially at times. (Path 15 has since been substantially upgraded.)

Even after the adjustments made in light of scheduled imports over the interties and in the day-ahead and hour-ahead markets, control room operators were compelled to make last-minute changes and purchases so as to balance generation and load in real time. Electricity demand may have been higher or lower than anticipated before the hour, transmission lines could have tripped, or power plants could suddenly go off-line. The shortfalls had to be accounted for in the real-time-imbalance market, also known as the spot-market (console 10). The real-time market was run by the real-time grid resources coordinator, commonly referred to within CAISO as the BEEPer. Each hour was divided into six ten-minute Balancing Energy and Ex-Post Price (BEEP) intervals in which an incremental price was set to encourage market parties to either increase (inc) or decrease (dec) their power output. Bids were submitted for inc-ing and dec-ing, and it was the responsibility of this real-time BEEPer to maintain that bid stack and process the inc or dec bids on the request of the generation dispatcher. (BEEPers have long since been replaced by software discussed in Chapter 6.)

Last but not least, there are consoles 11, 12, and 13, three positions that continue to be crucial to control room reliability. The CAISO generation dispatcher, commonly known as the gen dispatcher (console 11), managed the grid in real time by estimating how much inc-ing or dec-ing was needed to control the Area Control Error (ACE), which shows the relative balance between generation and load in California's grid. Maximum fluctuations in the ACE are set by the aforementioned WECC reliability criteria. It was the task of the generation dispatchers to keep the imbalances of the CAISO grid within these bandwidths. How well the generation dispatcher did his or her work, given the constraints of load and generation under which they operated and the resources they had available in the BEEP stack (the list of bids in the BEEP market prioritized according to price) and through the interties, determined the number of violations CAISO faced.

At the time of our research, the most important standards to assess the reliability of CAISO system operations were two control performance standards (CPSs), set by NERC.

As noted earlier, the first control performance standard (CPS1) compares the relationship of the frequency of the grid to the ACE. For example, if the frequency is above 60 Hertz and the ACE is positive (that is, the control area has more generation than load and is "overgenerating"), then the system operator is contributing to the frequency problem and further driving up the frequency of the system. On the other hand, if the frequency is above 60 Hertz but the ACE is negative (that is, the control area has more load than electricity), the operator is trying to push the frequency problem back and contributing to the stability of the system. This makes CPS1 a reliability tool for the behavior of system operators in the interest of keeping the stability of the grid. However, the strict CPS1 control standard is difficult to assess and monitor in real time. Violations usually are not known until after the events take place in the control room.

The second control room performance standard, CPS2, is much more directly visible in CAISO's operations. It makes use of the ACE as the crucial variable to monitor and limit excessive unscheduled power flows that could cause instabilities in the grid. CPS2 demands that the gen dispatcher minimize the fluctuations of the ACE within a maximum bandwidth every ten minutes. The standard requires a monthly 90 percent performance rate within these bandwidths, and allows CAISO a daily quota of fourteen ten-minute CPS2 violations. If CAISO's monthly averages fall below 90 percent, the regional reliability organization (WECC) imposes a fine.

Besides the CPS violations, CAISO's grid operations are constrained by disturbance control standards (DCSs). A frequency disturbance is logged when an instantaneous change occurs sending the frequency above 60.05 or below 59.95 Hertz. Path violations, in turn, record the crossing of safety limits (measured both in load and in time) that have been set for the different transmission lines. Because the paths represent vital links between the different grids, path violations are considered major violations and are penalized by the

WECC. Also, it is important to maintain key voltages along the entire grid, as described earlier.

The gen dispatcher and BEEPer worked closely together as a team. The inherent instability of electricity demand and behavior of the electricity system required constant fine-tuning and hands-on management of balance and load and therefore constant recourse to the real-time imbalance market. Next to the gen dispatcher was another gen dispatcher (console 12), who assisted the primary gen dispatcher and served as a backup for communications, monitoring, and assessment.

The grid operations shift manager (now shift supervisor, console 13) completed CAISO's control room team. The shift manager was responsible for overall coordination within his or her crew and acted as liaison for that team with parties outside the control room. During each twelve-hour shift, each manager had the ultimate control room responsibility for managing the bulk transmission grid. All critical decisions, including issuing emergency stages, were to be taken by the shift manager in consultation with his or her superiors.

Other departments within CAISO are critically related to CAISO control room operations. At the time of our 2001 research, the following departments constituted the control room's "wraparound," that is, direct support units performing vital tasks that allowed the CAISO control room to function reliably: Scheduling, Operations Support and Training, Market Operations, Operations Engineering, and Outage Coordination.

The Scheduling department was responsible for the process called prescheduling, the preplanning of generation and load schedules on the basis of the long-term and day-ahead markets. Operations Support and Training provided the personnel to back up control room operators in times of emergencies and threatened system crashes. They were also responsible for the training of control room personnel.

Market Operations designed and controlled the software for the different electricity markets in which CAISO played an active role at that time. Changes in market rules, CAISO's tariff (its governing legislation), responses to strategic behavior, and institutional changes resulting from the electricity crisis all had to be introduced into new software. Consequently, updates in market software occurred on a regular, almost weekly basis during the California electricity crisis in 2001. (Sometimes the new software developed in response to the crisis necessitated unavoidable gridwide experiments, such as in the introduction of a proxy bid system discussed in the next chapter.)

Operations Engineering undertook studies of the vulnerabilities and operating characteristics of California's electricity grid. Using computers to model parts of California's grid under specific operating conditions (nomograms), operations engineers were to anticipate and analyze the system's behavior under specific conditions. In emergency situations and periods of heavy system loads, operations engineers routinely entered the control room to provide technical support for specific parts of the electricity grid.

Finally, Outage Coordination enabled CAISO to have more influence and control over the planned and unplanned generation outages in California's restructured industry. Outage Coordination helped distribute the maintenance of power plants throughout the year, which in turn enabled control room personnel to better cope with controlling the grid.

All these support units were subsequently reorganized under the CAISO 2006 realignment, which had a variety of impacts on the wraparound's ability to support the control room. These are discussed in Chapter 4.

THIS WAS THE SYSTEM WE FOUND when we showed up in CAISO's control room in April 2001. We—a joint team then from Mills College in Oakland and Delft University of Technology in the Netherlands—undertook the bulk of our investigations between April and December, 2001. Sixty interviewees were identified and interviewed for an hour or longer: thirty-three in and around CAISO's control room in Folsom, California; eight in and around the Transmission Operations Center and the Operations Engineering units of Pacific Gas & Electric (PG&E) in San Francisco; five with a large generation supplier and private market energy trading dotcom (a senior generation official and control room operators in one of its large California plants); and fourteen others, including interviewees in the California Governor's Office, California Public Utilities Commission (CPUC), California Energy Commission (CEC), Electric Power Research Institute (EPRI), Lawrence Berkeley National Laboratory (LBNL), University of California, Berkeley (UCB), and Stanford University. More details on the research methodology can be found in Appendix 1. By far, most of our time and attention were directed at the California ISO, especially in and around its control room.

Our visits to the control rooms of CAISO, as well as of PG&E and a private generator, allowed firsthand observation of the interconnectedness of the state's restructured electricity network. There was not then one operator in the CAISO control room who was not closely connected to the outside through multiple communications and feedback systems. Everyone, all the time, used the telephone; pagers were ubiquitous; internal computers inside the control rooms "talked to" external computers; the AGC (Automatic Generation Control) system connected the CAISO generation dispatcher directly to private generators; the ADS (Automatic Dispatch System) connected the dispatcher directly to the bidder of electricity; dynamic scheduling systems in CAISO connected to out-of-state generators; all kinds of telemetry measurements from across the grid came back to the control room in real time; Web pages used by CAISO, PG&E, and private generators carried real-time prices and information; and on and on.

It is important to understand the nature of this network beyond the CAISO control room. In 2001, the gen dispatcher had a touch monitor speed dial that connected him or her directly to others in the distribution utilities and private generators. One connection was to PG&E's control room for its own market trading activities. The PG&E control room operator, in turn, had speed dial contact with the CAISO generation dispatcher.

Some of the plants the PG&E control room had formerly contacted directly were now owned by private energy suppliers. In our interview with plant operators in one of these privately owned plants, an operator showed us his direct lines, particularly one to the company's trading floor. These links come full circle in that, although CAISO's gen dispatcher did not have a direct line to the plant we visited, he or she did have one to that supplier's trading floor, just as did the operator in the plant's control room. In addition, we learned that the CAISO gen dispatcher might also call plant control operators informally (even though such calls were discouraged by new proprietary market relationships) during extreme peak demand days. Phone calls by operators who were wired to each other through such sophisticated technologies and software became the lifeblood of California's restructured electricity network.

BECAUSE THE RESTRUCTURED NETWORK was fragmented but interconnected, reliability in these circumstances was often heavily dependent on the CAISO control room's ability to pull resources to balance load just in time for the start of the next hour. Because of the time pressure this brought with it, operators could not rely completely on their specialized tasks and procedures but initiated a great deal of lateral communication to quickly and constantly relay all kinds of information necessary to "keep the bubble" with respect to the variables that needed managing given the performance conditions they faced. As a gen dispatcher put it to us in 2004, "As long as you keep the big picture, things are good. . . . You have to look broadly to see how a change has affected you. Stability of attention is very important—you can't lose focus." Another gen dispatcher put the point this way for his job: "It's the massive amount of multitasking, you got to be analyzing what's moving, how fast can it move, you've got to have a good overall picture of what's going on, all this simultaneously."

Operators not only had to respond quickly to unpredictable or uncontrollable events, they had to make sure that their responses did not exacerbate the balance problem, especially as confusion over what was actually happening was intense during the electricity crisis. In periods of confusion the risk is that one perturbation amplifies into another.[6]

What was most visible to us in 2001 was what we did not see: very few "normal" days. The electricity crisis had speeded up everything. More days became "peak" days. Table 2.1 shows the time breakdown by percentage one shift manager gave for the various activities he undertook during "normal" and "peak" days.

As the shift manager moved from normal to peak conditions, coordination requirements increased, while "firefighting" emerged as its own noteworthy requirement. Such transitions were common for the other operators in the

TABLE 2.1.
Percentage time breakdown of shift manager tasks

Normal Day: Activities and Time Spent	Peak Day: Activities and Time Spent
Routine 30%	Coordination on floor 30%
Procedure comments and reviewing 20%	Firefighting 25%
Coordination on the floor 20%	Dispersing info to departments 15%
Meetings 15%	Conference calls (plus prep time) 10%
Normal department communications 10%	Issuing operating orders 10%
Reviewing outages 5%	Routine and logging 10%

CAISO control room.[7] A BEEPer describes normal and peak day activities as follows:

My calls are with the SCs [energy schedulers], maybe three or four calls on a slow day, but on days when we're busy, the phone rings off the hook. On the whole how many people do you have calling you? It's hard to say, sometimes ten people calling in an hour; or in a slow day it is three or some calls. . . . When you get to know them over time, there's some that have more expertise, and there're others less skilled.

Other interviews at that time describe much the same increased tempo and pace in activities during the electricity crisis. It "can go from sheer boredom to sheer panic in one phone call," said a CAISO manager of real-time scheduling. Operators shared more information, paid more attention to more variables, and overheard more of what was going on around them. It became increasingly important during peak conditions for operators to have an overview of what was going on with respect to systemwide service and grid reliability. A new piece of information potentially affected everyone else in the control room when balancing load and generation. During peak periods, operators did not have a way to discriminate individually the important and the unimportant. This led, as we observed, to the need for increased information sharing, and informal, lateral, and collegial relations came to the fore, stretching into the wraparound and outside the CAISO control room into the rest of the high reliability network of distribution utilities and private generators.

During these peak times, wraparound staff from the support units started showing up or were called into the control room to help out with their specialized expertise (for instance, someone who "knows a lot of the San Francisco grid") or as a backup to an overworked control room dispatcher. Much of this was captured in the comments of a senior CAISO control room official about control room behavior during a Stage emergency in 2001:

The engineers are the resident experts in situations like yesterday [when CAISO had a Stage 2 emergency]. We augment the [control room] staff, giving more manpower to deal with the situation. Workload increases dramatically. Some of the desks get really active. The fishbowl [the conference room looking onto the control room] also changes, it becomes the emergency operations center. We try to improve communications to the outside in addition to helping grid operations. We try to keep lines of communication open to give good info. The same happens with the peak day call to PG&E and utilities. It's a tool for me to get instant updates with their boss and cut through their bureaucratic staff of operations. . . . In the stages, we augment [control room] staff to necessary levels. Engineers expert in overloading and resident experts on Northern California come out onto the floor. Some guys have expertise in particular areas, and it's good to have people to make the right decisions. You need a good support staff in this business. You are not going to know everything, and you need the rest to do that for you. When I get in trouble, they get the engineers. They have this intuitive ability to know when they're needed on the floor. I don't have to call them.

Our observations paralleled those of the official. During one 2001 emergency, as CAISO moved into a Stage 1 on its way to Stage 2, we observed the control room team expanding. The supervisor of the shift managers came out and started to help the shift manager. Wraparound staff who had worked the BEEP desk also arrived to help the BEEPer on duty. The then WSCC security coordinator began to walk around, talking more to the shift manager and the gen dispatcher. Specialist engineers augmented the scheduling and transmission operators. Operators and others on the floor were constantly scanning the computer monitors or wall monitors—for example, the shift manager was watching what was happening to reserves, with the gen dispatcher constantly watching the ACE. As they moved into Stage 2, the CAISO senior control room official came out on the floor. He knew a great deal about out-of-state generators. In the view of one informant, if there were to be deals made over the interties, he had the authority and connections to make that work. We were told, "He also knows the system inside out. He's . . . another special pair of eyes." The more information and the more expertise interpreting it, the lower the risk of operator error in such situations, other things being equal.

Some days were worse, when control room operators felt backed into a corner, where doing one thing could end up making something else worse. Complaints about "quick fixes" and "band-aids" and "firefighting" were common during these times. If conditions worsened further, a Stage 3 declaration to shed load (in other words, forced outages leading to "blackouts") was invoked. Eventually we saw the markets stop functioning altogether. "I was here, working as a new gen dispatcher, when I saw the market collapse. From one day to the other there were no more bids coming [into the real-time imbalance market]," a member of the CERS purchasing team told us. The utilities couldn't pay the prices being offered. "The destruction of the utilities' creditworthiness and the resulting responses by suppliers shattered all vestiges of a normal market," is how two analysts characterized the situation (Mensah-Bonsu and Oren 2001, 2).

Yet the lights stayed on by and large through all of this. How? This takes us back to our original research questions and the theories they engage.

CHAPTER 3

A FRAMEWORK FOR
HIGH RELIABILITY
MANAGEMENT

A S WE AND OUR COLLEAGUES OBSERVED
and interviewed still more in the CAISO control room,
we began to draw out from the mass of material a framework for understanding how the control room kept the lights on.

To return to the primary research question: How did this restructured network of control rooms, that operated within organizations and systems having different mandates and interests, actually ensure the provision of reliable electricity, even during the California electricity crisis? Our answer: The focal organization, CAISO, balanced load and generation in real time (that is, in the current hour or for the hour ahead) by developing and maintaining a repertoire of responses and options in the face of unpredictable or uncontrollable system volatility produced either within the network (such as by generators acting in a strategic fashion) or from outside (for example, by temperatures and winds).

Our research focus became the match between, on the one hand, CAISO's options and strategies to achieve its reliability requirement (namely, balancing load and generation and staying within reliability standards set for key transmission lines or "paths," while meeting the other parameter constraints) and, on the other, the unpredictable or uncontrollable threats to fulfilling the reliability requirement. A match results from having at least one option sufficient to meet the requirements under any given condition. At any moment, the possibility of a mismatch exists between the system variables that must be managed to achieve reliability requirements and the options and strategies available for managing those variables.

The match between option and system requirements can be visualized in the following way. Within the restructured network, the core reliability tasks are located in CAISO's main control room. It is the only unit that simultaneously has the two reliability mandates: keeping the flow and protecting the grid. Meeting the dual reliability mandates involves managing the options and strategies that guide actions of the independent generators, the energy schedulers (SCs), and the distribution utilities. As the focal organization, the CAISO control room deploys options that are networkwide; for instance, outage coordination is the responsibility of CAISO but involves the other partners in the restructured network.

In brief, CAISO high reliability management can be categorized in terms of the variety of network options that control operators have available (high or low), the volatility of the California electricity system (high or low) that they face, and the resulting performance modes adopted in the control room (Figure 3.1).

System Volatility

		High	Low
Network Option Variety	*High*	Just-in-time performance	Just-in-case performance
	Low	Just-for-now performance	Just-this-way performance

FIGURE 3.1. CAISO performance modes

Volatility is the degree to which the focal organization, CAISO, faces uncontrollable changes or unpredictable conditions that threaten the grid and service reliability of electricity supply—that is, that threaten the task of balancing load and generation. Some days are low in volatility. An example of high volatility was those days when a large part of the forecasted load had not been scheduled through the day-ahead desk, which meant for CAISO that actual flows were unpredictable and congestion would have to be dealt with at the last minute. Volatility refers to system instability, not to price movements alone.

Options variety is the amount of network resources, including strategies, available to CAISO with which to respond to events in the system in order to keep load and generation balanced at any specific point in time. The resources can be approximated with conventional grid parameters, including available operating reserves and other generation capacity, and available transmission capacity. High options variety means, for instance, that the grid has more than the required resources available (there are high or wide margins), low options means the resources are below requirements and, ultimately, that very few resources are left (low or tight margins).

The two dimensions together set the conditions under which the CAISO control room has to pursue its high reliability management. The conditions demand, and we have observed, four different performance modes for achieving reliability (balancing load and generation), which we term just-in-case, just-in-time, just-for-now, and just-this-way. *Low* and *high* are obviously imprecise terms, though they are used and recognized by many of our CAISO interviewees. In practice, the system volatility and options variety dimensions should be better thought of as continua without rigid high or low cutoff points.

Just-in-case performance (redundancy and maximum equifinality). When options are high and volatility low, "just-in-case" performance is dominant because of high positive redundancy. Reserves available to CAISO are large, excess plant capacity exists at the generator level, and the distribution lines are working with ample backups, all much as forecasted with little or no volatility (again, unpredictability or uncontrollability). More formally, redundancy is a state in which the number of different but effective options to balance load and generation is high relative to the market and technology requirements for maintaining this balance. Different options and strategies can achieve the

same balance. The state of high redundancy is best summed up as one of maximum "equifinality"; that is, there are many means to meet the reliability requirement, "just in case" they are needed. (Note that "equifinality" does not
mean that all options are equal in all respects, only that they are roughly
equivalent in meeting the balance requirement.)

In just-in-case performance, the control room reaches its maximum specialization and individualized competence. Sharing of information and perspectives is minimized, with integration of tasks done primarily through procedures, as in a classic coordination-though-task design. Each operator is doing
his or her job—performing according to formal rules established in advance.

There is no need to share information extensively or to have an integrated,
comprehensive picture of the entire network in order to balance load and generation. Control room operators do not need the real-time support of wraparound specialists or people to swap places with them or back them up. In
ideal terms, it's a normal day; in organizational terms, control room operators
have slack or redundancy.

Under performance conditions of low system volatility and high network
options, much of the wraparound remains nonactivated, though staff adjacent
to the control room are still developing software, testing new protocols, and
troubleshooting. Wraparound staff may take a break and visit control room
operators, and meetings are scheduled involving staff and operators. The big
risk in just-in-case performance is that someone is not paying attention when
they should be to potential changes in system volatility or network options variety. "What you want to do is avoid complacency," a senior utilities control
room official said about his operators. Problems are more likely to arise, in
other words, when "nonactivated" becomes inactive or "deactivated." "One of
my people," the control room official continued, "made a mistake because he
was complacent, too confident in what he was doing."

Just-in-time performance (real-time flexibility and adaptive equifinality).
When options and volatility are both high, "just-in-time" performance is dominant. Option variety to maintain load and generation remains high, but so is
the volatility of system variables, in both markets (for example, the 2001 rapid
price fluctuations representing strategic behavior by market parties) and technology (such as sagging transmission lines during unexpectedly hot weather).
This performance mode demands real-time flexibility; that is, the ability to

utilize and develop different options and strategies quickly in order to balance load and generation. Operators in the control room are in constant communication with each other and others in the wraparound, options are reviewed and updated continually, and informal communications are much more frequent. Flexibility in real time is the state in which the operators are so focused on meeting the reliability requirement and the options to do so that more often than not they customize the match between them; in other words, the options are just enough, "just-in-time."

More formally, the state of real-time flexibility is best summed up as adaptive equifinality: alternative options are developed or assembled as required to meet the reliability requirement. Substitutability of options and strategies is high for just-in-time performance, in which the increased volatility of the task environment is matched by the flexibility in options and strategies for keeping performance within formal reliability tolerances and bandwidths. As one ISO control room shift manager put it, "In this [control room] situation, there are more variables and more chances to come up with solutions." "It's so dynamic," said one of CAISO's market resource coordinators, "and there are so many possibilities. . . . Things are always changing."

Operators in this performance mode do not only have to respond quickly to unpredictable and uncontrollable events. They also must ensure that their responses are based on understanding variables so that these responses do not exacerbate the balance requirement, especially as confusion over what is actually happening can be intense at these times. However, the fact that the volatility is high focuses operator attention on exactly what needs to be addressed and clarifies the search for adequate options and strategies. People from the wraparound are pulled into operations and real time to extend the capability to process information quickly and synthesize it into a "bubble" of understanding covering the many more variables and complex interactions necessary in just-in-time performance. The activation of the wraparound and the expansion of the control room team under just-in-time performance conditions are largely directed to support error reduction and decrease the risk of operator misjudgment. In the just-in-time performance mode the beginnings of crisis management become visible (Boin, 't Hart, Stern, and Sundelius 2005).

Just-for-now performance (maximum potential for deviance amplification). When option variety is low but volatility is high, "just-for-now" performance is dominant. It is the most unstable performance mode of the four and the one that control room operators want most to avoid or exit as soon as possible. Operators commonly describe what they are doing as "firefighting," "band-aids" and "quick fixes." Options to maintain load and generation have become notably fewer and increasingly insufficient to what is needed in order to balance load and generation. This state could result from various conditions related to the behavior of the complex system that is the California energy sector. Unexpected outages can occur, load may increase to the physical limits of transmission capacity, and the use of some options can preclude or exhaust other options, for instance, using stored hydro capacity now rather than later. In this case, unpredictability or uncontrollability has increased while the variety of effective options and strategies is diminished. Now undertaking an option can actually increase volatility; for example, mitigating congestion on one transmission line may make things more volatile on another transmission line. As one WSCC security coordinator put it to us in 2001, using band-aids to fix the technology did not solve things, but led only to more band-aids.

More formally, just-for-now performance is a state summed up as one of maximum potential for "deviance amplification." Even small deviations in elements of the market, technology, or other factors in the system can ramify widely throughout the system (Maruyama 1963). Marginal changes can have maximum impact in threatening the reliability requirement; the loss of even a low-megawatt generator can tip the system into forced blackouts. A declaration of a Stage 1 or Stage 2 emergency (public alerts based on reserve generation capacity falling below 7 and 5 percent of current load respectively) may compel a CAISO senior manager to go outside official channels and call his counterpart at a private generator, who agrees to keep the unit online "just for now."

From the standpoint of reliability, this state is untenable. Here operators have no delusion that they are in control. They understand how vulnerable the grid is, how few the options are, and how precarious is the balance; they are keeping communication lines open to monitor the state of the network. They are not panicking and, indeed, they still retain the crucial option to reconfigure the electricity system itself, by declaring a Stage 3 emergency. "There is always an answer," a gen dispatcher told us. "Ultimately you can shed load. Doesn't mean though you want to do it."

When options become few and room for maneuverability boxed in (that is, when load continues to rise while new generation becomes much less assured and predictable), control operators become even more focused on the looming threats to balancing load and generation. As options deplete, wraparound staff in the control room come to have little more to add. There is less need for lateral, informal relations. Operators even walk away from their consoles and join the others in looking up at the big board on the side wall. "I'm all tapped out," said the gen dispatcher on the day we were there when CAISO just escaped issuing a Stage 3 declaration. Operators and support staff are waiting for new, vital information because they are out of other options for controlling the ACE themselves.

Just-this-way performance (zero equifinality). In this last performance mode for balancing load and generation, system volatility can be lowered to match low options variety. This performance mode occurs in the California electricity system as a short-term emergency solution. One major means to tamp down volatility directly is the hammer of crisis controls and forced network reconfigurations. The ultimate instrument of crisis management strategy is the Stage 3 declaration, which requires interruption of firm load in order to bring back the balance of load and generation from the brink of just-for-now performance. The effect of a Stage 3 declaration is to reconfigure the grid into a system under command-and-control management. Load reductions can then be ordered from major wholesale electricity distributors.

More formally, just-this-way performance is a state best summed up as one of zero equifinality: whatever flexibility could be squeezed through the remaining option and strategies is forgone on behalf of maximum control of a single system variable, in this case load. The Stage 3 declaration has become both a necessary and sufficient condition for balancing load and generation, again, by reducing load directly. This contrasts significantly with the other three performance modes. In those, options and strategies are sufficient, without being necessary. Under just-this-way performance conditions, the decision to shed load has been taken and now information is centered on compliance. The vertical relations and hierarchy of the control room extend into the restructured network, even to the distribution utilities in their rotating outage blackouts. Formal rules and procedures move center stage, including the making and ending of the Stage 3 declaration. Here the restructured network looks most like the structure it was before the deregulation, but only because of the exceptional requirements of grid reliability.

The operational differences we observed in the CAISO control room and the larger restructured network under these four performance modes are summarized in Table 3.1.

When we returned to CAISO in mid-2002 to present the results of our research—this being a year after our initial control room observations during

TABLE 3.1.

Features of performance modes for high reliability management

	Performance mode			
	Just-in-Case	*Just-in-Time*	*Just-for-Now*	*Just-This-Way*
Volatility	Low	High	High	Low
Option variety	High	High	Low	Low
Principal feature	High redundancy	Real-time flexibility	Maximum potential for amplified deviance	Command and control
Equifinality	Maximum equifinality	Adaptive equifinality	Low equifinality	Zero equifinality
Operational risks	Risk of inattention and complacency	Risk of misjudgment because of time and system constraints	Risk of exhausted options and lack of maneuverability (most untenable mode)	Risk of control failure over what needs to be controlled
Information strategy	Vigilant watchfulness	Keeping the bubble	Localized firefighting	Compliance monitoring
Lateral communication	Little lateral communication during routine operations	Rich, lateral communication for complex system operations in real time	Lateral communication around focused issues and events	Little lateral communication during fixed protocol (closest to command and control)
Rules and procedures	Performing according to wide-ranging established rules and procedures	Performing under prior and active analysis; many situations not covered by procedures	Performing reactively, waiting for something to happen	Performing to very specific detailed procedures
Orientation toward Area Control Error	Having control	Keeping control	Losing control	Forcing command and control

the peak of the 2001 electricity crisis—we were informed that well over 85 percent of control room activity was still in the real-time reliability performance modes of just-in-time and just-for-now. Price volatility was no longer the problem it had been in April 2001, but now new problems were arising from the aftermath of electricity restructuring. The reported distribution of time spent across the performance modes (Figure 3.2) turns out to be especially important, as we will see in Chapter 4, when the dominant performance mode shifts away from just-in-time.

To summarize: CAISO's high reliability management in 2001 consisted of the ability to (1) maintain the balance of load and generation within any one of the four performance modes of options variety and system volatility; (2) adapt to externally generated shifts in system variables so as to maintain the balance of load and generation; and (3) move out of "just-for-now" performance mode as soon as possible into "just-in-time," "just-this-way," or "just-in-case" performance modes. In short, grid and service reliability are high when the balancing of load and generation can be sustained across all four performance modes. Thus *the cross-performance maneuverability of operators in response to shifting volatility and options conditions proved to be as important to high reliability management as balancing load and generation in any one performance mode.*

That is how the lights stayed on. Surprisingly, current theories about high reliability suggest that what we found is not possible.

	System Volatility	
	High	*Low*
High	Just-in-time performance −70%	Just-in-case performance −10%
Low	Just-for-now performance −20%	Just-this-way performance <1%

Network Option Variety

FIGURE 3.2. Estimated percentage of time spent in each performance mode, June 2001–May 2002

CHAPTER 4

WORKING IN PRACTICE
BUT NOT IN THEORY[1]

T HE LITERATURE AND THEORIES ABOUT high reliability in large technical systems range over diverse disciplines, and it is clear that significant gaps in that literature must be accounted for if we are to understand the high reliability management witnessed in CAISO during and since the turn of the century.

To make our case, we first examine reliability theory and writings about the management of complex technical systems, beginning with quality control and human factor approaches. We then address two specific theories for the operation of large technical systems and the long-standing debate between them: high reliability theory and normal accident theory. Because neither theory accounts for the level of high reliability management we observed, we conclude the chapter with a further development of our own reliability framework.

Interestingly, some literature that would seem especially pertinent proves to have little to say about high reliability management. The literature on the economics of deregulation, with only very few exceptions (for example, Hawk 1999 on a transaction costs analysis of California deregulation proposals) maintains a willful silence, if not studied indifference, to how reliability across deregulated units is actually to be managed—except to say that we get what we pay for. One would also expect, because most of this nation's critical infrastructures are privately owned, that a great deal of thought and work would have been directed to reliability management in the private sector.[2] A growing literature on "business continuity" has emerged, devoted to telling business organizations how they can better anticipate disruptions or respond better if and when they occur. But little attention seems to be given to the idea that reliability may require more than prior anticipation and preprogrammed emergency response (a topic we explore in Chapter 8).

INTEREST AMONG THEORISTS in organizational reliability—the ability of organizations to manage hazardous technical systems safely and continuously even when the risks of error are high—has grown dramatically in recent years (LaPorte and Consolini 1991; Rochlin and von Meier 1994; Schulman 1993a, 1993b; Perrow 1999 [1984]; Roberts 1993; Sagan 1993; Sanne 2000; Weick 1987; Weick, Sutcliffe, and Obstfeld 1999; Beamish 2002; Evan and Manion 2002; Weick and Sutcliffe 2001; Perrin 2005). Several reasons account for the heightened attention and concern.

First is the intensifying social dependency on "high performance" hazardous technologies ranging from nuclear weapons and power to large jet aircraft, medical technologies, complex electrical grids, and telecommunication systems. Many large technical systems impose relatively tight error tolerances on operators and maintenance personnel, and the consequences of the errors are high, often ramifying beyond the user to bystanders and society at large (Perrow 1999 [1984]). In addition, concern for reliability among organization theorists has grown because of major, high-profile accidents so commonly known as to be recognizable from a single phrase, such as Exxon Valdez, Three Mile Island, Bhopal, Challenger, Columbia, and Chernobyl. Many accidents illustrate vividly that technical design alone cannot guarantee safe and continuous performance (Weick 1993; Vaughn 1996; Meshkati 1991). Finally, the blowback from 9/11 and Hurricane Katrina has intensified interest and concern over critical infrastructures and their reliability in the face of potential terrorist attack and catastrophe (National Research Council 2002).

Despite the upsurge in interest, the analytics of reliability research in organization theory are deeply rooted. Some of the earliest research appears in quality control (QC) analysis. QC has been centered on ensuring reliability in both organizational products and production processes. One quality control historian traces the beginnings of the field and its focus on identifying and averting failure in production to medieval guilds, with their emphasis on long training, formal testing of skills, and careful inspection of task performance and products (Juran 1986). Statistical QC began in the 1920s and helped isolate causal factors, and their precursors, in production errors and failures (Duncan 1986; Ott and Schilling 1990).

A related approach to reliability research has been human factors analysis (HFA). Led by psychologists (as opposed to the engineering roots of much quality control research), HFA focuses on the impact of physical designs and task requirements on human performance (Salvendy 1997; Perrow 1983; Norman 2002; see also Wagenaar 1996). As general tendencies, QC seeks to achieve reliability by controlling worker behavior to match task requirements, while the HFA is directed toward securing reliability by crafting strategic organizational and task variables to human requirements. High reliability management, as we shall see, requires mutual adjustments along both dimensions.

As important as quality control and human factor approaches to organizational reliability have been for analysts and organizations, they differ from more recent reliability studies. Research during the 1980s and 1990s did not address production reliability *per se*. For some organizations, errors, accidents, and failures undermining safety affect much more than production. They may jeopardize the survival of the organization itself and its members as well as significant numbers of people outside the organization.

The earlier research treats production reliability as a marginal or fungible property whose costs could be traded off against other organizational values such as efficiency, speed, and product performance (Schulman 2002). The reliability analysis of the 1980s and 1990s was quite different. It dealt with error and failures that have far-reaching, often unacceptable implications for safety, not just inside but also outside the organization. This is not reliability as a probabilistic property that can be traded off at the margin with other organizational values, but reliability directed toward a set of events whose occurrence must, as nearly as possible, be deterministically precluded (Schulman 2004). (Recent "crisis management" literature is about what happens when those events are not precluded ([Boin, 't Hart, Stern, and Sundelius 2005].)

The differences between the marginal reliability of earlier production approaches and the more recent focus on precluded-events reliability are summarized in Table 4.1.

The beginning of a new orientation to organizational reliability can be traced to works by James Reason (1972) and Barry Turner (1978) that connected human and organizational factors as systematic producers of major technical failures. A key work in this approach has been Charles Perrow's *Normal Accidents* (1999 [1984]). Perrow added a new dimension to his earlier work on technology and organizations (Perrow 1979). He categorized technologies along the dimensions of complexity and tight coupling. For Perrow (1999 [1984]), tightly coupled systems are highly time-dependent in not allowing for delays or unexpected contingencies; fairly invariant and inflexible in terms of the sequence of activities required; and characterized by little slack and few resources available to tolerate delays, stoppages, and the unexpected when they do occur. A system is complexly interactive if it has unfamiliar, unplanned, or unexpected sequences of activities that often are not visible or immediately comprehensible.

TABLE 4.1.
Marginal and precluded event reliability

Variable	Marginal Reliability	Precluded Event Reliability
Context	Efficiency	Social dread
Risk	Localized	Widely distributed
Calculation	Marginal (variable cost)	Nonfungible (fixed requirement)
Standards	Average or run of cases	Every last case
Learning	Trial-and-error learning	Formal learning with limited trial and error
Orientation	Retrospectively measured	Prospectively focused

In the view of Perrow, these complexly interactive and tightly coupled technologies are particularly problematic from the standpoint of organizational reliability. They pose the likelihood of "normal accidents"—founded on the risk factors embedded in their design—irrespective of what strategies organizations adopt for their management. These technologies are accidents waiting to happen and capable of changing their conditions or states with a speed and interaction that defy the real-time reactions of operators or the anticipation of designers and planners.

Perrow's framework set a limiting condition for the organizational reliability of large technical systems that has resonated through subsequent studies (Sagan 1993; Evan and Manion 2002). Consistent with that perspective, the history of electricity and the drive for greater reliability are punctuated by large-scale, widespread blackouts—East Coast, West Coast, and in-between—whose collapses and near misses compelled technological change to improve the reliability of the grid (for example, Hauer and Dagle 1999).

Countering Perrow's normal accidents theory (NAT) is the work of a group of researchers who identified in case studies what they found to be a set of "high reliability organizations" (HROs). These organizations (a nuclear aircraft carrier, nuclear power plant, and air traffic control center) established comparatively excellent performance records in managing technologies of high complexity and tight coupling (LaPorte and Consolini 1991; Rochlin and von Meier 1994; Roberts 1993; Schulman 1993a). They were found surviving in highly unforgiving political and regulatory niches with respect to errors and failures. They were able to do so, it was found, because of structural, managerial, and cultural factors that buffered the organizations from the hazards of tightly coupled and complex technologies and, in effect, mitigated the risks these systems posed. Among the factors observed by the HRO researchers are those summarized in Figure 4.1 (a fuller discussion of the features is found in Appendix 2).

The managers of high reliability organizations begin with a clear specification of core events that simply must not happen (LaPorte 1996). To this they add the specification, through careful examination of experience and causal analysis, of a set of precursor events or conditions that could lead to core events that pose unacceptable hazards to the organization and beyond. The

- High technical competence
- High performance and close oversight
- Constant search for improvement
- Hazard-driven adaptation to ensure safety
- Often highly complex activities
- High pressures, incentives, and shared expectations for reliability
- Culture of reliability
- Reliability is not fungible
- Limitations on trial-and-error learning (operations within anticipatory analysis)
- Flexible authority patterns under emergency
- Positive, design-based redundancy to ensure stability of inputs and outputs

FIGURE 4.1. Principal features of high reliability organizations (from 1980s research)

precluded and precursor events bound an "envelope" of reliability within which these organizations seek to operate.

HROs develop elaborate procedures to constrain behavior and task performance within the envelope. At the same time, the organizations feature a "culture of reliability," that is, a widespread, shared mindfulness toward conditions that might causally lead to error and failure (Weick and Roberts 1993; Langer 1989). Careful identification and formal specification of core precursor events enable the HROs to avoid operating in areas where reliability or "safety" is being jeopardized (Schulman 1993b).

High reliability organizations are also characterized by the simultaneous pursuit of contradictory or paradoxical properties of reliability (Rochlin 1993). Error protection regimes that guard against one type of error (say an error of omission) are likely to make another type of error (errors of commission) more likely. As such, HROs must clearly specify operational procedures and standards and yet cannot become insensitive or inattentive to the unexpected. In short, they have to pursue simultaneous strategies of anticipation and resilience. Another contradiction to be managed arises between formal design principles and the actual operational demands of case-specific responses. High reliability organizations are able to buffer these paradoxes in their day-to-day operations.

Another feature discovered in the research on HROs is the ability to transform formal roles, reporting, and authority relationships under emergency conditions or increased stress. Typically this means bypassing formal hierarchy and the development of lateral, less formal, and more flexible modes of communication and coordination (Roberts 1990). Much HRO work is carried out in teams, with great emphasis on the cultivation of high levels of technical competence through personnel selection and training throughout.

High reliability organizations recognize that key organizational properties such as attention, close coordination, and mutual trust across units that have to rely on one another are not constants and cannot be treated as givens. They are subject to decay and have to be continually renewed to the high levels required in the organizations. Routines can numb mindfulness (Langer 1989); shared understandings can erode. It is not invariance in these properties but rather the attention to and careful management of their fluctuations that helps promote the high reliability organization (Schulman 1993a).

Finally, high reliability organizations exist in environments that share an intense aversion to the events they are managing to preclude. This means HROs are carefully watched and regulated, and are constrained from internal drift or goal displacement away from high reliability commitments by the constraints imposed by the wider environment. At the same time the environment supports the organization in treating reliability as nonfungible; that is, it generally insulates the organization from pressures to trade off reliability with other attributes (such as costs) under intense market competition. Reliability is a fixed requirement without any real-time substitutes. Thus, for instance, electricity consumers absorb the security and reliability costs of nuclear power plants, and all airlines are required to practice similar maintenance procedures under close FAA regulation. This regulation and support allows HROs to incorporate redundancy in technical designs and to invest a great deal in anticipatory and contingency analysis.

WHAT VAUGHAN (2005) HAS CALLED the "great divide" between NAT and HRO approaches to understanding reliability continues to persist. Whether the HRO features constitute a sufficient or even necessary set of conditions for preventing normal accidents, that is, precluding unacceptable events, remains unanswerable. Although the dispute between normal accident theory and high reliability research continued through the 1990s (Perrow 1994; LaPorte 1994; Rijpma 1997; Weick, Sutcliffe, and Obstfeld 1999), in its most extreme form the dispute has centered on an assertion that is unfalsifiable.

No amount of good performance can disprove Perrow's viewpoint concerning normal accidents because it can always be said that an organization is only as reliable as the first catastrophic failure that lies ahead, not the many successful operations that lie behind. For example, Perrow insists that there have not been more nuclear accidents on the order of Three-Mile Island because "we have not given large plants of the size of TMI time to express themselves" (Perrow 1999 [1984], 12). High reliability theorists can never fully explain high reliability to the satisfaction of NAT proponents, and yet we have high reliability no matter how impossible it is in theory. As Gilbert Ryle, the philosopher, famously put it, "Efficient practice precedes the theory for it" (Ryle 1949, 30).

At the same time, more subtle problems underlying NAT have largely gone unacknowledged by its proponents. One dimension, which Perrow asserts to be an independent variable, tight coupling, has significant ambiguity surrounding its identification and understanding. It can be difficult, for example, to distinguish tight coupling as a cause or a consequence of failure. In massive flooding across mid-Western states that occurred in July 1993, water flows overwhelmed dams and reached levels so high that a set of spillways across several states, which had been considered independent state flood protection devices, became tightly coupled (Hey and Phillipi 1994). The failure of physically separate dams and spillways to contain unprecedented water levels was really the independent variable that turned a loosely coupled set of elements into a tightly coupled system. At best we could ascribe tight coupling as a latent feature of such spillways, a feature that follows upon a specific magnitude of failure.

The HRO research perspective has its own conceptual and empirical difficulties. The research has centered on a small number of selective case studies at a single slice in time for each organization. These few cases do not constitute a proven argument that the features identified in the organizations were truly necessary ones (Schulman 1993b). Further, high reliability organizations research has in some respects asserted high reliability as a defining characteristic rather than a performance variable of its organizations. This leaves unanswered the question of which features, if any, and in what amounts or combinations, can contribute to higher reliability (along a continuum) in organizations. Fortunately, more recent research has begun to broaden the analytic focus on reliability from structure to process in organizations, especially the cognitive and sense-making skills and strategies of their members (Weick, Sutcliffe, and Obstfeld 1999; Sanne 2000; Weick and Sutcliffe 2001; see also Hodgkinson and Sparrow 2002).

BOTH NORMAL ACCIDENTS THEORY and high reliability research centered around HROs are in agreement in one respect, however. Each predicts that the restructuring of the California electricity sector should have noticeably undermined the reliable provision of electricity.[3] For its part, NAT would see the coupling of the PG&E and SCE grids into a reconfigured statewide grid, with altogether novel and higher flows of energy, as an increase in the technology's tight coupling and complex interactivity. The probability of cascading failures should increase accordingly.

High reliability theory would have come to a similar conclusion, but by a different route. HROs, including some of the older integrated power utilities (see LaPorte and Lascher 1988), are almost exclusively preoccupied with ensuring stability of internal processes within single organizations, which, if not closely managed, greatly magnify consequences for error and failure. Consequently, they seek to stabilize both inputs and outputs, much as manufacturing firms might seek a vertical integration of upstream and downstream elements in production and marketing in order to stabilize the manufacturing process.

High reliability research would predict that deregulation of the electricity sector would necessarily introduce substantial volatility in electricity provision by increasing the unpredictability of inputs and lessening the applicability of prior anticipation and analysis, thus undermining the principal features associated with high reliability performance identified in Table 4.1 and Figure 4.2. Restructuring has brought unbundled generation, and a far more complex system for the transmission and distribution of electricity, in which control room operators are now dispersed through organizationally distinct units, with the plant generator linked to the company trading floor, the transmission center of the utilities dedicated to distribution only, and CAISO as the manager but not owner of the grid. What was once a single organizational culture of reliability is now a network of divergent interests bordering on polarized at times.

Yet instead of widespread unreliability, our 2001 research found a high reliability network of deregulated generators, utilities, and transmission managers that was remarkably reliable even during the extreme California electricity crisis. Neither HRO-research nor NAT would have forecast anything like the high reliability performance we observed. How to explain the gap between theory and practice?

CAISO INTERVIEWEES TOLD US IN 2001 that while avoiding grid burnouts and major service interruptions—that is, while maintaining the commitment to precluded-events reliability— CAISO had not one official operational reliability standard that had not been pushed to its limits and beyond in the California electricity crisis. The emerging reliability criteria, standards, and associated operational bandwidths had one common denominator: they reflected the effort by members of the restructured electricity network, particularly the CAISO control room, to adapt reliability criteria to meet circumstances that they could actually manage, circumstances that were increasingly real time in their urgency. What could not be controlled "just-in-case" had to be managed "just-in-time;" if that did not work, performance had to be "just-for-now;" or, barring that, "just-this-way" by shedding load directly. In each instance, reliability standards became the bar that the operators could actually jump.

The standards at issue were many. Not only were operating reserve limits questioned, but the system operated reliably—in particular, peak-load demands were met continuously and safely—at lower reserve levels than officially mandated in WSCC (now WECC) standards. CPS criteria were disputed and subsequent efforts sought to change them. There was mounting pressure to empirically justify standards that were formulated *ex cathedra* in earlier periods and had since become "best operating practices." One senior CAISO engineer told us during the electricity crisis,

Disturbance control standards (DCS) say that if I have a hit, a unit goes offline, ACE goes up, I have to recover within ten minutes. Theory was that during that time you were exposed to certain system problems. But who said ten minutes? God? An engineering study? What are the odds that another unit will go offline? One in a million?

So now with WSCC we have turned the period into fifteen minutes, because the chance of another [unit] going offline is low, as we know from study of historical records. . . .

There is, in fact, a paradox between having reliability standards and having multiple ways to produce electricity reliably. On the one hand, the standards are operationalized (in terms of WSCC) and, because performance can be empirically gauged against these operational measures, they become the chief measuring stick of whether electricity is being provided reliably. On the other hand, the standards were being redefined in the electricity crisis, because only by pushing the standards to their limits and sometimes beyond were the lights kept on. A senior manager in the CAISO operations engineering unit, responsible for a large body of procedures, told us "part of the [control room] experience is to know when not to follow procedures. . . . there are bad days when a procedure doesn't cover it, and then you have to use your wits." A former control room official summed up this operator orientation nicely in 2004: "procedures are always with the caveat, 'in your best judgment.'"

The best way to understand the distinctiveness of this approach to reliability is to compare and contrast the principal features of traditional HROs (for example, the Diablo Canyon nuclear power plant) with the approach taken in CAISO high reliability management, both of which seek and achieve high reliability in connection with electricity.

Table 4.2 displays a multiple-dimensioned continuum in high reliability service provision between two approaches to reliability management for two tightly coupled, complexly interactive technologies in our electricity infrastructure. Note that one important feature of the high reliability management of CAISO is the large proportion of that management that occurs in real time under conditions of high system volatility. This is a departure from the large preponderance of anticipatory, just-in-case and just-this-way management found in much of the earlier HRO research. Indeed, the California system cannot be reliable with respect to grid or service reliability without having the options to perform just-in-time or just-for-now. What this preoccupation with real-time reliability means is that the CAISO control room and others in the restructured network are engaged in a very different kind of reliability management than found in much of the earlier HRO research.

TABLE 4.2.

Comparison of selected features of HRO and HRM

Traditional HRO Management (HRO Research on Diablo Canyon)	High Reliability Management (Research on California's Restructured Electricity Network)
• High technical competence	• High technical competence
• Constant search for improvement	• Constant short-term search for improvement
• Hazard-driven adaptation to ensure safety	• Often hazard-driven adaptation to ensure safety
• Often highly complex activities	• Often highly focused complex activities
• Reliability is nonfungible	• Reliability is nonfungible in real time, except when service reliability jeopardizes grid reliability
• Limitations on trial-and-error learning	• Real-time operations necessitating improvisation and experimentation
• Operation within anticipatory analysis	• Operations outside analysis
• Flexible authority patterns within HRO	• Flexible authority patterns within focal organization and across HRM
• Positive, design-based redundancy to ensure stability of inputs and outputs	• Maximum equifinality (positive redundancy), adaptive equifinality (not necessarily designed), and zero equifinality, all depending on network performance conditions
• Low input, process, and output variance	• High input and low output variance requires high process variance

Also, the Table 4.2 differences in approaches to trial-and-error learning and to redundancy are particularly important to understand. The earlier literature on HROs found that they avoided anything like the large-scale experimentation we found in the California restructured electricity network. The improvisational and experimental are what we have summed up as adaptive equifinality in just-in-time performance. What was to be largely avoided in the HRO has become the *sine qua non* for service and grid reliability in the restructured network, with its real-time preoccupation.

There were experiments on the large scale—that is, on the scale of the California grid as a whole—that were gridwide because CAISO could not do otherwise in 2001. They were undertaken involuntarily. CAISO had little choice, for example, in introducing "proxy marketing" software over the whole grid. Over the course of the day it was introduced, complaints were made that price information was wrong, numbers were not showing up, and the information wasn't in real time.

Why experiment this way? Because the status quo had become untenable for the system manager, thanks to increased volatilities introduced into the electricity system through the restructuring-induced energy crisis. Bids from market traders were not coming in, and operators had to take corrective action (we now know that some of the bids were being withheld for strategic reasons by private energy traders). The action that CAISO took was to create proxy bids for the remainder of generation capacity not bid in by the energy traders. In such ways, real-time operations and experiments become synonyms.

But even involuntary experiments can become occasions for learning, much as near misses are in other sectors (for instance, Comfort 1999). The real-time experiment, deliberate or inadvertent, becomes a design probe from which operators can learn more about the limits of service and grid reliability. "You don't learn as fast as you can, until you have to respond to something that requires fast response," argued a senior control room manager at PG&E's transmission operations center.

The traditional HRO nuclear power plant, such as Diablo Canyon, would never undertake (or in all likelihood be allowed to undertake by regulatory agencies) such experiments, as it seeks both stable inputs and stable outputs. In the restructured network's case, there is no stable resting point for the grid and its demands. There were few routines and operating procedures in 2001 that could stabilize inputs (such as load, generation availability, or grid conditions) in order to facilitate highly reliable outputs. One reason why operations-as-experimentation continues is that shedding load is a live option for maintaining real-time reliability. If just-in-time performance fails, operators can always shift into just-for-now or just-this-way performance modes. Reliability in this case requires large *process variance* in order to match volatile inputs to regularized outputs.

OPERATORS MANAGING THE ELECTRICAL GRID are likely to be confronted with uncontrollable or unpredictable input variance. Weather, load fluctuations, and generation or line outages vastly increase the input variance operators are likely to encounter. Yet electricity must be always on and at all times load must balance generation on the grid.

There are so many system components and variables (many beyond the control of control operators and dispatchers) that the electrical grid can assume a huge variety of system states. Given the wide variation of component combinatorics (line conditions, generator outages, and temperature and weather conditions, as well as load levels) there is scarcely a modal or "normal" configuration for the system. That is why high reliability depends on managing load and generation across multiple performance modes as and when needed. This is also why skills in pattern recognition and the formulation of new action scenarios are so important in high reliability management.

When high input variance must be reduced to low output variance it requires high process variance; that is, a wide range of options by operators and controllers.[4] The process variance includes options assembled as needed in real time with significant improvisational behavior by operators (for example, adaptive equifinality in just-in-time performance). This process variance is quite different from that encountered at a conventional high reliability organization such as a nuclear power plant.

It therefore should not be surprising that a related finding from our California case study is that both complexity and tightly coupled interactions can serve, and often do serve, as positive sources of high reliability performance. This observation challenges the perspectives of both the early NAT theory and HRO research. But complexity actually allows adaptive equifinality in just-in-time performance because many variables are in play, such that some combination of options or their components can be stitched together at the last minute. The tight coupling in fact positions operators within CAISO such that they can implement their reliability solutions as well as exercise the Stage 3 option of command and control.

An important aspect of this real-time reliability is the substitutability of options and components. Substitutability is key for just-in-time performance, when system volatilities are compensated for by flexibility in response. There are many ways to skin the cat. This view contrasts with that of normal accidents theory. For NAT, the ability to substitute elements is a property of loosely coupled systems and linearly interactive ones. According to Perrow (1999 [1984], 96), "What is true for buffers and redundancies is also true for substitutions of equipment, processes and personnel. Tightly coupled systems offer few occasions for such fortuitous substitutions; loosely coupled ones offer many." Yet we found that the same substitutability is core to just-in-time performance, when the technology of the grid remains as tightly coupled as it has always been.

A primary reason why tight coupling and complexity were found to serve as a resource for options in the CAISO case study but not in the earlier Diablo Canyon case lies in the differing roles played by causal analysis. Near-complete anticipatory analysis is central to the high reliability performance of traditional HROs; not so for the restructured network of the California grid. The urgency of real time makes it crucial to "read" feedback in terms of signature events (such as shifts in the ACE or frequency) that can guide the balancing of load and generation, in the absence of operators having full understanding of the system in the process.

None of this is to say that tight coupling and complex interactivity are not a problem for California's electricity infrastructure as it currently exists. They are, but in a very different way than NAT proposes.

ACCORDING TO NAT, tightly coupled and complexly interactive technologies are particularly hazardous because each element summons up a contradictory management strategy. Tight coupling entails, according to Perrow, a requirement for centralization of authority and operations, with unquestioned compliance and immediate response capability. Complex interactivity, in contrast, requires decentralization of authority to cope with unplanned interactions on the ground by those closest to the system.

But who reconciles the need for anticipation and careful causal analysis with the need for flexibility and improvisation in the face of turbulent inputs into complex and tightly coupled systems? In our research we found a crucial, albeit neglected role for middle-level professionals—controllers, dispatchers, technical supervisors, and department heads—in doing the balancing act so necessary to the real-time reliability of these networked systems.[5] We term them *reliability professionals* in recognition of their overriding commitment to the real-time reliability of their systems, and the unique set of skills they bring to their tasks. We have talked in the preceding chapter about the importance of their performance modes and their ability to maneuver between modes in order to manage for high reliability. It is now time to extend that framework further and discuss how and in what forms the match between skills and tasks takes place.

In our framework, the quest for high reliability in tightly coupled, highly interactive critical infrastructures can be characterized along two dimensions: (1) the type of knowledge brought to bear on efforts to make an operation or system reliable, and (2) the focus of attention or scope of these reliability efforts. The intersection of these dimensions forms a cognitive space for high reliability management as sketched in Figure 4.2.

The knowledge base from which reliability is pursued can range from formal or representational knowledge, in which key activities are understood through abstract principles and deductive models based on these principles, to experience, based on informal or tacit understanding. In this way, the knowledge base varies in terms of the mix of induction and deduction used in reliability management.

At the same time, the scope of attention in reliability management can range from a purview that embraces reliability as a stream of predictable system outputs, encompassing many variables and elements, to a case-by-case focus in which each "case" is viewed as an event with distinct properties or characteristics. Scope is typically articulated as the different scales, ranging from specific to general, to be taken into account in the provision of critical services.

Let us now turn to key "positions" within this cognitive space. Each is a different mix of skills and perspectives along the two continua. Toward the extreme of both scope and principles is the macro-design approach to reliability.

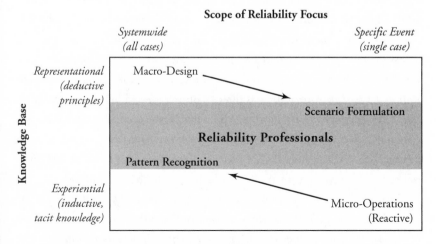

FIGURE 4.2. Cognitive space for CAISO reliability professionals

Here formal deductive principles are applied to understanding critical processes. It is considered inappropriate to operate beyond design analysis, while
the analysis is meant to cover an entire critical system, including every possible
case of performance. Design in this sense is more than analysis; it is a major
control instrument of system behavior. This approach dominates operations
in a nuclear power plant, where operating "outside of analysis" constitutes a
major regulatory violation.

At the other extreme of the cognitive space is the activity of continually reactive behavior in the face of real-time challenges at the micro-operations
level. Here reliability resides in the reaction time of system operators working
at the event level rather than the anticipation of system designers for whatever
eventuality. The experiences of crisis managers and emergency responders are
exemplary in this regard.

The macro and micro extremes at either "corner" are, however, insufficient
for ensuring reliability. Designers cannot foresee everything, and some design
errors are inevitable. (Indeed, the more complete a set of design principles attempts to be, the harder it is to ensure that the full set won't contain two or
more principles contradicting each other.[6]) On the other side, case-by-case reactions by their very nature are likely to give the operator too specific and individualized a picture, resulting in the operator losing sight of the garden for
the flowers. Experience in micro-operations can become a "trained incapacity"
that leads to actions undermining reliability, because operators may not be
aware of the wider ramifications of what they are doing and experiencing.

What to do then, if the aim is reliability? Clearly, mixing perspectives
across the reliability space horizontally from one corner to the opposite corner
is unlikely to be successful. A great deal of research has found that attempts to
impose large-scale formal designs directly onto an individual case—to anticipate, fully deduce, and determine the behavior of each instance from systemwide principles alone—are very risky (Perrow 1999 [1984]; Majone 1978;
Turner 1978). At the same time, reactive operations rooted in individual experience hardly constitute a template for scaling up to the system level.

Instead of horizontal connections, Figure 4.2 indicates that reliability is enhanced when *shifts in scope are accompanied by shifts in the character of the
knowledge base.* In this perspective, being reliable requires more and different
knowledge than is found at the extremes of *a priori* principles and individual

operator experience. Given the limitations of the extremes in the cognitive space, it becomes important for reliability that operations take place in positions closer to a shared center by (1) moving from "firefighting" reactions in a single case to recognizing patterns across cases and (2) moving from systemwide designs to formulating scenarios for action contingent on localized situations and conditions. We know from research that reliability is enhanced when managers, technical personnel, and operators apply designs less globally and relax their commitment to a set of principles that fully determine system operations. This happens when they embrace a wider range of contingencies in their application of design and customize scenarios to fit the case at hand. From the other direction, reliability is enhanced when operations shift away from real-time firefighting and "quick fixes" toward the recognition of patterns and trends across cases in the system and the imputed consequences of these for system reliability. The direction of the arrows in Figure 4.2 indicate both the factors that push those concerned with reliability from the extremes of macro-design and micro-operations and the factors that pull them to pattern recognition and scenario formulation as a way of improved reliability management.

It is in the middle ground between scenario formulation and pattern recognition that we find the domain of reliability professionals. CAISO control room operators and their support staff are reliability professionals because they are the ones with the unique skills and knowledge base that keep the California grid reliable. They are neither system designers nor narrowly focused reactive operators. They excel in identifying patterns and trends and in formulating scenarios and protocols to cover contingencies. They translate recognized patterns and contingent scenarios into reliable service provision.

This middle ground is not easy to find, nor is it guaranteed to exist even in an organization that says it takes reliability seriously. It is difficult to move beyond macro and micro perspectives on a large technical system, to bring together logic and experience and—perhaps even more difficult—theory and practice. It is, however, in this middle ground where systemwide patterns and case-specific scenarios can be reconciled and where individualized perspectives, be they at the operator or designer levels, can be transformed into unique networked knowledge for reliability management. In this sense the domain of the reliability professionals is a crucible for the continual generation of new knowledge covering the reliable operation of a complex technical system.

To summarize, reliability professionals operate in a cognitive space driven by pattern recognition (the process of "sizing-up" a situation that connects a set of events or occurrences to a model or schema) and a process of formulating scenarios for action under diverse contingencies (protocols that cover the range of variance in system behavior evidenced in individual cases or events).

The rest of this book is about what those specific skills in pattern recognition and scenario formulation are and how they operate. Here however it is important to underscore a more general finding: again and again we have found in our research that these professionals and their distinctive cognitive domain are neglected by system designers, economists, regulators, and the public alike. Yet it is with this group and their professionalism that we believe the greatest gains in grid and service reliability are to be found. The stakes have never been higher in getting both theory and practice right.

The next step in explaining high reliability management lies in better understanding these professionals, their work, and the cognitive skills they bring to bear on it. To do that requires us to bring the story of CAISO up to date.

CHAPTER 5

CAISO IN 2004
Control Room Reliability Endangered

FTER THE BULK OF THE 2001 INTERVIEWS, we and our colleagues wrote up our findings. We returned to CAISO in mid-2002 to confirm our impressions, and visited CAISO periodically thereafter. In an April 2003 visit we were told that the control room was spending some 60 percent of its time in real-time reliability (just-in-time and just-for-now performance modes). The improvement from 2001, in other words, was slight, even though another year's stabilization efforts had passed (including a "must-offer" requirement placed on generators and substantial new generation capacity coming on line).

Our intention was to continue tracking how control room time spent in real-time reliability was changing as CAISO evolved and matured as an organization. When we sat down in mid-March 2004 with CAISO's then chief operating officer for a progress report on how things had been going recently, he focused on a major load-shedding event that had occurred on March 8, 2004, which was said to have been induced by CAISO control operators. This event disturbed him deeply. Blackouts had taken place in southern California. An internal CAISO review found fault with the two generation dispatchers on duty in its control room at the time. (The title of a subsequent CAISO press release read, "Preliminary California ISO Internal Investigation Finds Operator-Error Contributed to 20-Minute Outage in Southern California.")

The COO (now VP-Operations) spent considerable time describing to the two of us how concerned he and others were about the event. He had trouble describing how the control room was that day when he walked in after the event. "I can't explain it any other way than it was dead," he said. "It was dead in the control room." He asked if we could follow up on why that event had happened.

His request fit in with our growing interest in refining the notion of high reliability management in the control room. We wanted to sharpen our understanding of the cognitive dimensions of what CAISO control room operators and support staff actually did in their performance modes and within their professional domain of competence.

As we explored operator skills and capacities and the challenges operators confronted under a variety of grid conditions, we were continually struck by how important to reliability the match was between operator skills and the actual real-time task requirements of balancing load and generation to reliability. The match kept being described in connection with an edge. Operators talk about being "near the edge" or "on the edge." As one shift supervisor put it, "Not only are inexperienced operators working in unfamiliar conditions, but experienced operators sometimes don't know how close they are to the edge."

We began seeing the need and usefulness of having indicators of when the operators approached or went over this cognitive edge. These indicators would measure operating conditions or challenges that are tied to high reliability management and could potentially identify conditions that significantly affect measurable reliability outcomes. If there were such indicators, they could help CAISO to identify factors that threaten the balance between skills and tasks, and quite possibly develop strategies to keep operators away from these conditions. Research into other high reliability organizations, as discussed in Chapter 3, reveals that they develop "precursor strategies" designed to keep operating conditions away from clearly specified circumstances that could degrade or exceed operator skills (Schulman 1993b; Carroll 2003).

Precursors are indicators of when reliability professionals have moved to the edge of their performance capabilities. They signal when operators are approaching a threshold at which their cognitive skills are no longer matched to the task requirements they face. The precursor zone is where the skills of the professional to recognize patterns and formulate scenarios are significantly challenged. As such, the precursor zone lies within the cognitive domain of reliability professionalism but at its edge, as shown diagrammatically in Figure 5.1.

We proposed in mid-July 2004, and the COO agreed, that we would undertake a new intensive phase of our research. As before, we would be focusing on the match between skills and tasks, a "reliability envelope" in the control room. Given continuing performance challenges at CAISO, the best way to do this, we felt, was to look for indicators of the edges of this envelope: What are signs and signals that a specific skill, state of mind, or frame of reference among reliability professionals is being overloaded or degraded? If it were possible to identify more clearly factors defining the reliability edges of the real-time operational envelope of the crew, then preventive and protective strategies might be devised. We began a round of intensive interviews—but this time almost exclusively in the CAISO control room and among immediate support staff—that have continued up to the time of writing (2007).

After starting in mid-2004, it gradually dawned on us that something was wrong in the control room. Our uneasiness began to interrupt the indicators

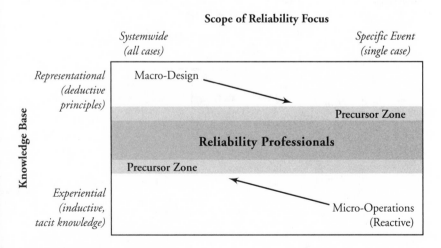

FIGURE 5.1. Precursor zone

research. We heard in our interviews more frequent mention of the role of "luck" in maintaining balances in the grid during critical periods. As one CAISO engineer put it to us in September 2004 about a recent high-load day, "The stars were aligned . . . and we lucked out with the rest of the western grid." Such luck should scarcely be a major factor depended upon in high reliability organizations.[1]

As we discussed what we were seeing and what our interviewees were actually saying to us, we realized over the first few months of the new research phase that CAISO's high reliability management was in jeopardy and in a way different from the California electricity crisis. What had been the dominant performance mode of just-in-time in 2001 and afterward (Figure 5.2) now looked to us in mid-2004 to have shifted to just-for-now, with a serious degradation in the maneuverability of operators to just-in-case and just-this way modes (Figure 5.3).

Plainly speaking, operators were describing more ways they were getting hammered; it was less clear to them just what success in real time meant; second-guessing was on the rise; and the "next time around" seemed to be coming sooner rather than later—all the signs to us of just-for-now performance.

We began to realize that operations seemed to be coalescing around what operators themselves had identified to us in 2001 as the most unstable performance mode they faced. Once this became clear, we suspended our research to write a special report to the chief operating officer. We expressed our con-

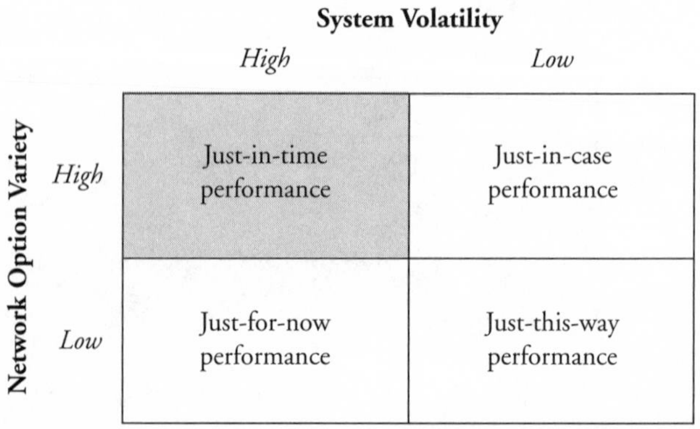

System Volatility

		High	Low
Network Option Variety	*High*	Just-in-time performance	Just-in-case performance
	Low	Just-for-now performance	Just-this-way performance

FIGURE 5.2. 2001

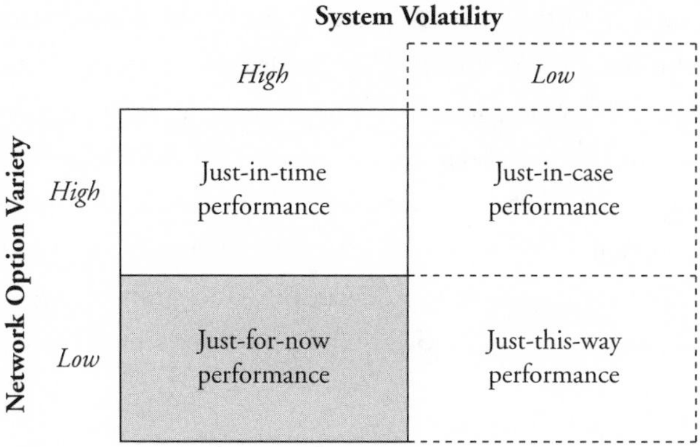

System Volatility

	High	Low
High	Just-in-time performance	Just-in-case performance
Low	Just-for-now performance	Just-this-way performance

Network Option Variety

FIGURE 5.3. 2004

cern over what we were observing, and the fact that the framework for high reliability management discussed in the preceding chapters actually indicated just how threatening the changes we were seeing were. This chapter presents that report and discusses its impact. The next chapter brings the CAISO story up to date with new changes made in response to the concerns highlighted in our report as well as the latest developments in CAISO's evolution.

MORE THAN HALF A CENTURY AGO, political scientist Harold Lasswell, along with others, laid out a framework for the policy sciences (Lerner and Lasswell 1951). Lasswell's chapter, "The Policy Orientation," set ambitious aims. The policy sciences should be focused on the big issues; add to our knowledge through what has become a problem-oriented, context-sensitive, and multimethod approach (for example, Brunner and Willard 2003); focus on not just policymaking but also policy execution; and recognize that any item—seminar, record, or report—can contribute to the field, even if outside conventional social science.

Lasswell was very clear that the contributions for improved decision making are not limited to journal articles, published books, or official reports. "[W]here the needs of policy intelligence are uppermost, any item of knowledge, within or without the limits of the social [science] disciplines, may be relevant" (Lasswell 1951, 3). This includes "examining written records, and also . . . interviewing the participants (13)."

What follows in this chapter is the January 2005 report we delivered to CAISO's chief operating officer. We reproduce it here, with minor edits,[2] not only because it had an impact and followed from our framework for high reliability management, but also in the conviction that many advances in the understanding of high reliability management, both practically and conceptually, could well come from such sources. It is a lengthy report, but one that repays close reading as it summarizes key points, introduces core concepts developed in Part Two, and lays the groundwork for recommendations more fully developed in Part Three.

TO: CAISO Vice President, [COO] Operations[3]

FROM: Emery Roe and Paul Schulman

RE: Provisional report on findings and suggestions from control room research

DATE: January 7, 2005

This report updates you on our research findings as of the above date. Background information is first provided, our provisional findings are summarized, and we conclude with suggestions.

This phase of our research at CAISO has focused on the skills that enable control room operators to function as reliability professionals—to identify patterns and anticipate scenarios that are key to successful ongoing electric grid management. The skills include pattern recognition, focus and concentration, memory (both short term and long term), reaction time, situational awareness, communication, and decision-making skills.

The match between these operator skills and the task requirements across all performance conditions is a "reliability envelope" for control room operations. Our current research identifies mismatches between operator skills and task requirements as indicators of the "edge" of that reliability envelope. We have found major challenges in keeping operations at CAISO within that envelope.

We are observing that CAISO control room operators are "riding out" ongoing problems by being as reliable as they can in the circumstances they now find themselves. Those circumstances are, however in our view, the most unstable with respect to ensuring reliability of the grid.

I. BACKGROUND TO RESEARCH

In our July 2004 memorandum of agreement with CAISO and as part of our ongoing research into reliability management, we proposed to sharpen our understanding of what CAISO control room operators and support staff do in connection with the different performance conditions they face.

Our 2001 research identified four performance conditions within which control room operators needed to work in order to cope with all states of the grid they are responsible for managing. The conditions depended on the volatility of the system at the time of operation and the variety of options available to balance load and generation in the face of that volatility. That earlier table is reproduced here.

When operating reserves are high, excess capacity exists at the generator level, and transmission and distribution systems are working with ample backups, all much as forecasted with little or no volatility. Control operators are able to perform with a maximum of planning and anticipation (just-in-case performance).

Our earlier research, however, found that a great deal of the time your control room operators face performance conditions of higher system volatility, where reliability in balancing load and generation is achieved in real time. When options and volatility are both high, "just-in-time" performance demands

		System Volatility	
		High	*Low*
Options Variety	*High*	Just-in-time performance	Just-in-case performance
	Low	Just-for-now performance	Just-this-way performance

FIGURE 1. Four key performance conditions for CAISO

real-time flexibility, that is, the ability to utilize and assemble different options and strategies quickly in order to balance load and generation. "Usually there're options; there is not just one way to do things," a reliability coordinator said recently. "If I'm going to give suggestions, I give two or three."

We also found in our earlier research that when options are few but volatility remains high, "just-for-now" performance dominates. This is the most unstable performance mode of the four and the one that control room operators most want to avoid. Here they feel cornered, and they do not have the flexibility to maneuver. "Our hands are tied," one generator dispatcher put it. Temporary—and some not-so-temporary—band-aids, quick fixes, and gap-filling are required to keep the system going, "just for now." Last but not least, when options are insufficient to cope with high system volatility, a "just-this-way" performance mode was available at the time of our earlier research. This consisted of official Stage 1, 2, and 3 emergency declarations, which reduced volatility by command-and-control measures to shed load that led to controlled blackouts.

In our view, the right side of Figure 1 is no longer functioning in most CAISO control room situations. We have said that reliability depends on being able to operate in all four conditions, but we find that the operators we interviewed this time are almost exclusively operating on the left side in real-time reliability performance conditions.

You may remember we called control room operators and their support staff "reliability professionals." These professionals operate with initiative and unique knowledge bases that keep the system reliable. They are neither system designers nor narrowly focused reactive operators. Instead, they excel at recognizing systemwide patterns and anticipating more localized contingency scenarios about what to do when the unpredictable or uncontrollable happens. The figure from our earlier research describing these differences and the domain of the reliability professional is also reproduced for ease of reference.

We began the latest phase of research in July 2004 and have continued up to the present. Over twenty-six interviews with CAISO personnel across departments have been undertaken, including with control room personnel as well as individual support staff. On many interview days, we spent hours observing operations in the control room, including the transition on October 1

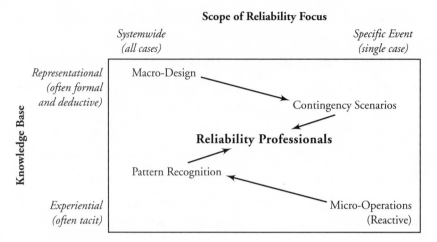

FIGURE 2. Reliability space for control room operators and support staff (reliability professionals)

to the RTMA [the Real-Time Market Applications software, replacing the BEEP desk, for dispatching last-minute energy in what were argued would be optimal ways]. We were also fortunate to attend the April–May 2004 CAISO summer training sessions.

II. PROVISIONAL FINDINGS

It is widely known that there is deterioration in concentration and memory reported by shift managers and generation dispatchers their first day back from vacation. In addition were sporadically mentioned problems in sleep patterns that carried over into the work next day, lapses of focus during the night shift, the "after-lunch lull," the reluctance of new recruits to share information among others in the shift, and the transition period between crew shifts.

That said, we have not found evidence of, nor do operators report, a significant increase in errors or lapses in the skill set of individual operators. We are aware of a decline in the experience base of new control room recruits, however.

While we have yet to find control operators drifting significantly to the edge of their reliability envelope in skill erosion or performance instability, we nevertheless do find that grid management, in our judgment, is closer to the edge of overall system reliability for four major reasons:

- The edge of the reliability envelope is moving closer in on the operators

- Successful grid management now requires that control operators work more closely to the boundaries of permissible (e.g., official NERC) reliability bandwidths and criteria

- The options for reliability management are themselves more volatile

- Staff support for reliability management has altered in character

The four factors have pushed grid management more frequently into the just-for-now performance mode (Figure 1). In terms of any theory of reliability with which we are familiar, this is the most problematic and least reliable set of performance conditions.

The four factors are examined in some detail, as our suggestions follow directly from the following analysis.

1. *The edge of the reliability envelope is closer.* Performance modes that were once available are really no longer available for current operations.

It is true that recent improvements have increased the control room's "command and control" over generation resources. Yet many real-time challenges in the control room do not allow operators the benefit of *just-this-way* performance (Figure 1). The emergency stage declarations are not helpful in stabilizing the grid volatility and increased loads that the operators now face. Moreover, the addition of new generation sources, which should have improved *just-in-case* performance, has increased system complexity and interactivity through congestion and mitigation problems, where more energy needs to flow across transmission lines than those lines can permit. In short, the right side of Figure 1 is being eroded away. The edge of performance conditions has moved in on the operators, who now work, it seems to us, almost exclusively under real-time reliability conditions of *just-for-now,* when not *just-in-time.*

How has this happened? Operators and their support staff have a clearer understanding of the limits and risks of their maneuvers for grid and service reliability than earlier. "Maybe ignorance was bliss before," a reliability coordinator[4] said. "We may never know how close we were to the edge then because we didn't know what was causing what we were seeing. Now we know that better." More so than in the past, we are being told that taking one action to improve reliability can make reliability worse somewhere else in the system. This is taking place, moreover, within a deregulated industry that is already working with fewer backups and reserves than the earlier vertically integrated utilities: "I know they had more margin to work with," said a generation dispatcher, "but we don't have that luxury. We're operating closer to the edge than they were."

2. *The changing reliability requirements for grid management.* Our earlier research found that there was an unavoidable overlap between operations and experimentation when it came to reliable operations of the grid. For example, at the height of the 2001 electricity crisis, CAISO had little choice but to introduce the experimental proxy marketing software gridwide. We were actually in the control room the day it was introduced, and we saw the need for its coming into operation across the entire grid then.

But it is one thing to experiment—that is, operate in "unfamiliar" conditions—when you have four performance modes to maneuver across. It is quite another thing to experiment when restricted to one or two performance modes on the left side of Figure 1. Reliability literature concludes that, while trial-and-error learning goes on in high reliability organizations, it does so in ways that deliberately avoid testing the boundary between system continuance and collapse. We observed that boundary being tested on the night of October 1, 2004, in the CAISO control room with the introduction of RTMA.[5] This was not just risk taking, it was risk making. Your operators were confident they could handle what happened, but not confident about what would happen. "Like I say," a reliability coordinator told us that evening, "when we're on the cutting edge, you can cut yourself."

Let us be clear about what we are arguing. CAISO was not the only ISO that was making a significant changeover that evening [of the RTMA changeover]. More to the point, a great deal of planning had gone into CAISO's RTMA changeover. As we also observed that evening, once the critical hours had passed, the shift manager started to encourage his generation dispatcher to formulate scenarios and ascertain patterns in how RTMA was actually working (two weeks later another shift manager told us he was encouraging similar behavior)—just what we would expect from reliability professionals. Further-more, in interviewing operators before and after RTMA's introduction, the clear consensus is that RTMA is working better than they had thought it would. All this is good news, but our broader point is that the changeover took place and is taking place in a control room whose options remain limited and whose performance conditions are now all but isolated to the left side of Figure 1.

The introduction of market software initiatives is taking place in circum-stances where successful grid management now requires that operators accept system conditions closer to allowable reliability limits. "There are more and more instances of single contingency situations than people see or think they see," reported one shift manager. Other operators told us of being increasingly one contingency away from loss of service.[6] Recent load increases to levels pro-jected for years ahead are another indicator of the challenge to keep within limits. The location of new generation away from the load requirements and the mitigation requirements that this and the forward contracts have entailed were repeatedly mentioned as constraining reliable grid management.

3. *The volatility of options.* Our earlier research characterized performance conditions of terms of the balance between system volatility and the variety of options to respond to that volatility. Options could be enlarged independently of what was happening to system volatility. From our recent observations, it appears that the options themselves have become more volatile, and signifi-cantly so.

Control operators speak of the risk of managing themselves "into a corner" by use of options. Ongoing, multiple mitigations are the primary example but not the only one. A senior control room official told us that thirty-four areas of interzonal congestion had been identified, i.e., "thirty-four balls you get to juggle at one time." Operators describe more situations in which they feel forced to work in "unstudied conditions." "Our basic foundation of reliabil-ity," a shift manager told us,

is that we operate the system to studied conditions, but now we're studying different conditions, and now some things are changing on a weekly basis . . . and we [control room operators] are not convinced nor are the engineers convinced that we've had ad-equate study for each of these configurations.

This too is taking place within a context where some options, such as automatic remedial action schemes (RASs), have long been recognized as potentially introducing greater volatility into the system. Indeed, it is difficult not to see RASs as an increasing part of just-for-now performance. "Everyone knows we install more RASs, that's the quick fix we put in," one operator said during the summer training session. "We don't believe most customers are aware of the fact that there's a higher chance of them being interrupted with RASs than they were before without them," a shift manager told us.

Much of the volatility that CAISO faces is a consequence of the increased complexity and tighter coupling of elements in the California grid. "We have the most complicated grid in the world," argues a senior operations engineer. In reliability theory, complexity and tight coupling are properties that challenge the ability of control operators to prevent cascading failures or errors. Indeed, cascading events are the limiting case for the skills of reliability professionals in terms of their ability to recognize patterns and anticipate scenarios (Figure 2).

Our earlier research found that complexity and tight coupling, in addition to their risks, actually functioned as a reliability "resource" for your control operators, particularly under just-in-time conditions. It allowed them to use the complexity of multiple generators, lines, and switching systems as the raw material to improvise remedies on a case-by-case basis for real-time grid management. Control operators were able to assemble a wide variety of solutions to many real-time grid management problems, in effect using one part of the system to compensate for a shortfall in another.

Now it appears that the grid retains all the risks of tight coupling and complexity without the positive features just described. Operator options themselves induce more volatility than we saw in the past. Mitigating for congestion in one zone raises problems in another. Rapid load changes can "unravel" a complex mitigation solution.

4. *Changes in support staff and functions.* In our earlier work, we were struck by the significant human capital available to support control operators under real-time performance modes (i.e., just-in-time and just-for-now). The major source of support was the technical advice provided personnel in what we call the "wraparound," those units literally based around the control room and including operations engineering, market operations, outage coordination, and information technology support, among others.

From inside the control room, the contribution of the wraparound, with the exception of a few stars, now seems more mixed. We have observed at times a considerably more problematic relationship between operators and support services. "Departments are writing contracts and completely tying our hands, taking away resources. They're solving a settlement issue and creating a reliability issue," complained a shift manager. A BEEPer adds: "One of the biggest problems that the back don't understand is that sometimes their expectations are set too high for us when it's busy out here—like my having to check prices, [which is] really difficult to do [then]; or creating software to help them when they are not really asking you what you need."

Earlier 2001 interviews recorded complaints about telemetry snafus, software crashes, and dropouts in displays. This time around, we have been hearing more about software upgrades that do not work as well as promised, workarounds for problems that are not rectified and, most prominently, procedures—too many procedures; procedures introduced only to be rewritten, clarified, or rescinded; so many procedures that operators can't keep up with them—yet in a context where operators know that procedures and experience are core to maintaining reliability. In the words of the BEEPer,

Five crews can't keep up with all these [procedure] changes; an email directive is not effective; this causes problems from shift to shift when you're stuck in certain rules but haven't communicated that to the others you're working with. You end up not communicating it right, sometimes you get somebody on your shift who is not speaking your language, is working off a different set of procedures. It's like working for different utilities sometimes.

Workarounds—"temporary" software to get around the glitches—are the clearest example of just-for-now conditions for control room operators whose "whole careers here have been around workarounds," said a senior control room official.

In other words, it is difficult not to conclude that the tools being refined by the wraparound are themselves introducing uncertainty into grid management, at least from the perspective of the control room operators. To the latter, the wraparound looks increasingly like a "back room" where office managers make decisions for the control room. A good number of the tools introduced since the California electricity crisis have improved performance—but even then the edge has moved inwards. Speaking of summer 2004, a senior control room official said: "I was just as scared this summer as I was in the [energy] crisis. The day we had the highest loads in southern California, the ties [transmission lines from other states] were all in, 50MW could have gone off, and we'd be done. . . . It was worse than the energy crisis, we were just right up to the edge [and] had seven system peaks." "We're at the edge," says an operations engineer of present conditions. "We are closer to the edge in terms of load growth, congestion management, but" a reliability coordinator adds, "we have better tools than we did four years ago." Even the good news of better tools sounds like running to beat your shadow.

What do these four factors add up to?

In our judgment, they indicate that changes in CAISO are increasingly serving to formalize grid management into the least reliable performance mode, just-for-now (Figure 1). Operators are increasingly locked into performance that is both at the edge of system reliability and closer to the edge of the reliability envelope of control operator skills. "There's a concern we're having to use ingenuity to get out of a lot [more] situations and sooner or later you are going to get boxed in," concludes a senior shift manager. The risk is exacerbated by two factors. First, this is occurring within the context of increased operator turnover and with senior management struggling to keep trained staff. Second, CAISO remains incompletely designed and organized. "There're not many startup organizations after seven years that have as much flux as we do," another shift manager told us. We turn now to a fuller discussion of the risks associated with prolonged just-for-now performance.

III. THREATS AND IMPACTS OF PROLONGED
JUST-FOR-NOW PERFORMANCE

The just-for-now performance mode, where control room options are few and volatility still high, is precisely the one where the mismatch between cognitive skills of the operators and the complicated task requirements is likely greatest, and where the risk of concatenating errors is the highest.

Just-for-now performance requires concentrated and extraordinary levels of attention by those concerned. "The complexity is such that we have engineers on twenty-four-hour call," reports an operations engineer. Yet, in the words of a shift manager, "to master attention to detail is not in the timelines we operate under." Accordingly, violations of control performance standards (CPS2) have gone up, stability limits are increasingly threatened, errors are amplified, markets remain hard to manage, and the SCs [security coordinators in the private sector buying and selling energy] keep pressing for more procedures. Managing the Area Control Error [ACE] within the regulatory bandwidths does not help out with the transmission line congestion problems, and events "which we've never seen before" become unavoidable. More formally, just-for-now performance can be summed up as the attenuation or absence of "bandwidth management" as the primary mechanism for control room management of the grid.

High reliability research finds that the key to maintaining reliability is the skilled management of inevitable fluctuations in performance parameters and conditions. In some high reliability settings the boundaries of these fluctuations—the reliability "bandwidth"—are fixed and discrete. They may be defined in formal procedures, embedded in technical models and operational protocols, or locked in by law or regulatory proscription. In other cases the boundaries are subject to negotiation among competing reliability mandates such as water flow, water quality standards, and endangered species legislation.

In either case, the high reliability organization is mindful of the borders of its bandwidth. Even when subject to negotiation and change, each change is noted and recorded. What we have observed in our CAISO interviews and control room observations is that the edges of the bandwidth are more uncertain, complex, incomplete, and less agreed-upon than during the time of our previous research. Where engineering analysis indicates that certain performance conditions are closer to the edge of system reliability than previously thought, there seems to be no formal hardening of the bandwidth to those conditions. What constitutes an "unfamiliar condition" itself seems subject to dispute between wraparound engineers and control room operators we have interviewed.

High reliability organizations work by identifying a set of core events that simply must not happen. In electrical grid management, for instance, these would include the collapse of the grid, damage to physical assets, and knock-on damage to adjacent control areas. Activities, tasks, and system design are all directed toward avoiding these events.

What is less appreciated is that these organizations also identify through experience and careful analysis a set of "precursor" events or conditions that could lead to the core events. For CAISO, these precursor events are CPS violations, quality of control deterioration, and one-contingency-away-from-shedding-load conditions as well as conditions in the control room that force operational error. These too are managed through procedures, protocols, and technical designs. Both core and precursor events are outside the permissible reliability bandwidth. In all cases, there is mindful agreement about the edge of the bandwidth.[7]

What we are seeing at CAISO is that more and more grid-management operations are conducted within the realm of precursor events. This is reflected in CAISO's own documentation of the deterioration in quality of control and the significant increase in CPS violations immediately upon the introduction of RTMA. "We weren't meeting the reliability criteria," the senior control room official said of that period. Many operators report they are more frequently in one-contingency-away conditions, with or without RTMA. "You're seven minutes into the stability overload," recounts one shift manager, "and anything could happen. Just one contingency and then who knows what is going to happen. The thing could cascade." These are highly risky situations, because when you do not manage the precursor events but you are still reliability seeking, then that very search for improvements (a key feature of high reliability organizations) can itself lead to more volatility and less reliability.[8] Even though RTMA represents an attempt to reduce the volatility around last-minute market changes as a way of improving grid management, its introduction was in the realm of precursor events that night.

Let us be very clear about this. We are not saying that any particular instance of operating in "unfamiliar" conditions is a degradation of reliability. When you see that the present set of options is not sufficient to prevent disaster and you have a possible option that could avert disaster but falls outside of the set of current options, you cannot automatically assume that action based on that analysis increases volatility or degrades reliability. In this example, the criterion of moving into unfamiliar areas assumes patterns and scenarios are still in place from which to make an assessment of the risks associated with failing to make a move. The point we are making in this report, however, is that operators are being pushed into situations where few or no patterns or scenarios are yet available or appraised. You have to be in studied conditions before you choose to move into "unfamiliar" conditions for the purposes of reliability management.

What this all means is that "bandwidth management" has been lost or diminished as the primary mechanism for control room management of the grid. But not everyone agrees that they are outside the bandwidths. "We do not operate in unstudied conditions," insist two operations engineers we interviewed. Yet, even though we were told several times during this round of interviews by those in the control room and the wraparound that "the operator always has the flexibility to keep the lights on," this begs the larger question: Is "turning the lights off" itself a volatile option in conditions that are unfamiliar?

In brief, we find that the reliability bandwidths are themselves the major unstudied condition in CAISO. Instead of bandwidths, we find quick fixes have become the order of the day because many problems require real-time correction and can't be postponed. "We have band-aids piling up on band-aids; there's a concern we're working our way into a death spiral, when there's not enough bodies to throw at a problem and control it," one shift manager reported. Even the dissenting view is not reassuring. A reliability coordinator told us: "We have more RASs, a lot of band-aids on the system, but there always have been the band-aids. I'm not sure it is any worse than it was; the engineers just have found more problem spots. But we are getting more of one solution leading to more problems. You solve that problem but something happens elsewhere because of it."[9]

When we looked at CAISO in our 2001 research, it was clear the control room operators were the frontline of defense in managing the system. They are now in a performance mode where it is less clear that they can serve as a first line of defense. In addition to being "tapped out of resources," operators also now complain of being "out of gas" at the end of their shifts. While some operators argue that performance standards are getting clearer and the consequences of not meeting them more severe, many operators, in our view, find themselves under conditions which force errors to happen. There are several reasons for this. First, the error happens unintentionally; e.g., because of a recent lawsuit against CAISO, fewer people are now on the floor because of mandatory lunch breaks. One Wednesday, the generation dispatcher goes to lunch and then what happens? "We're in a new paradigm," concluded a former generation dispatcher; "lunch break cost them [a major] violation" that day.

Second, forced error can be more a matter of choice, where operators recognize that they have to tolerate an error, such as a CPS2 violation, in order to prevent others. Another reason is that they are operating in conditions which undermine their abilities to recognize patterns and anticipate scenarios—in short, their skills and experience are least helpful. Operators "watch the weather all the time, but now it's lots of variables and the level of nervousness is higher," a senior member of market operations told us.

Moreover, there may be no patterns: "What gets me stressed is when something happens and I don't have an immediate answer," said a shift manager. "We got to do something and I am not sure what to do." Other patterns fade away. "It's hard to remember," one dispatcher replied when asked to compare the control room today with what it was like in the 2001 electricity crisis. "It all blends together." So too do things blend together in the short run. "My short-term memory has gone to hell," reported one shift manager:

I don't even know my phone number anymore; the numbers are always changing. You get to a mental point, why should I commit to memory generation ratings, transmission line ratings. I just go to the books now, because ratings have been in such flux, with upgrades, or RAS schemes, and ratings have a million variables that change. There is no hope to keep up, so subconsciously you don't keep up.

Even the long term gets foreshortened. "Short-term memory is important, long term is nonexistent," a generation dispatcher put it to us. "Ask me about yesterday and I don't know what I did." The risk, of course, is not knowing tomorrow what you know today.

What are the major manifestations and impacts of all this just-for-now?

First, it seems to us that CAISO control room operators no longer "own" the problems they have to manage, at least to the extent we saw in our earlier research. It is true that some operators "own" any given procedure that is being developed. We however are talking about something different.

The case studies on high reliability organizations demonstrate that operators own the problems they manage; they have control over the variables they are charged with managing and the conditions under which they manage.

Control operators feel they are authoritative with respect to their decisions, directives, tools, and procedures. "This desk is 100 percent decision making," said one generation dispatcher. "If you can't make a decision, you shouldn't be within a hundred feet of here," said another. This decisiveness, in the view of the reliability literature, includes both an acceptance and a full understanding of operator tools and tasks. "There is so much more done by computers, one of the things I like is to know what the logic is behind the thing," a generation dispatcher and former shift manager told us.

We find a significant lack of reliability ownership among control room operators in many dimensions, such as ownership of procedures and decisions, the conditions under which the control room functions, and the adoption of new tools and software. When shared problems are differently owned and managed, then all manner of complaints about second-guessing are to be expected. We have heard more complaints about "second-guessing" and "quarterbacking" in this stage of research than we did earlier.

Second, not only has ownership of problems diminished, in our opinion, the responsibility for solutions has also become more problematic. We were impressed in our 2001 research by how skillfully control room operators were able to sort out, pick, and assemble options just-in-time. Indeed, one of the pulls to real time was that operators took pride in the choices and improvisations they made under pressure. Now, operators feel much more pushed into just-for-now, moment-by-moment performance. The options set is more circumscribed and their "decisions" less a matter of choice than necessity. They are less and less the owners of their improvisations, because they feel less responsible for the situations in which they find themselves.

Third, not only are operators less accountable for the problems and solutions they manage, they also feel less claim to the reliability standards governing that management. We along with others have noted a marked decline in regard for CPS2 violations this time around. Our discussion above about band-aids and the increased operation in the realm of precursor events also applies here.

What this all means is that the space for reliability professionals in the middle of Figure 2 has contracted. Reliability professionals, such as control room operators, should be interested in ensuring the entire system is reliable from end to end, from system design to individual case. Instead, control room operators are being as reliable as they can be in the face of circumstances into which they are being pushed.

Fourth, because of the preceding changes, the nature of reliability management has itself changed in the control room. To reiterate, operators are increasingly, it seems to us, managing to the edge of grid failure, where "managing reliably" has come to mean managing to prevent real-time failure in grid or service reliability, *just for now.* What needs to be noted is the negative impact of this factor on the operator's professionalism.

This inability to use their skills fully (as they do in just-in-time performance) and the persistent demand to make due with gap-fillers and band-aids necessarily erodes professionalism and de-skills trained people who are now pushed to the edge of error for problems they no longer feel they can own and solutions they no longer consider theirs. "We have to adapt on a day-by-day basis," said one shift manager. "Yes, we get by day by day," said another shift manager to us, "but it's the long term I don't know." One shift manager, arguing that control room operators had "more complete" solutions five years ago, put the change this way:

> There's a lot of situations you don't feel you have the correct tools to handle the situations, situations where the outcome doesn't seem to be the right outcome. . . . You have more and more situations where the outcome you have to work for doesn't make anyone happy, where you have to put a lot of effort that no one is going to be happy with. We have grid management outcomes as well as policy that we shake our heads at; they become the product of thirty different competing interests, solutions that don't satisfy anyone. . . . There is a less complete and quality solution.

A generation dispatcher talked about there being fewer ways to "back out gracefully" from bad situations. Another dispatcher described it as "we rarely get in a situation where we want . . . there is usually so much moving." Yes, options are there, but not as many nor as varied as we saw earlier (Figure 1). None of this is helped by the widely recognized facts that (1) new control room recruits continue to work in situations of high flux but now have significantly less experience with all facets of electric power than the previous generation of dispatchers, while (2) control room operators are still expected to know everything about their position when worse comes to worst. Nor is this helped by a surprisingly increased skepticism about operator skills and accountability among personnel in the wraparound.

So what is CAISO to do? We believe it must get out of just-for-now performance conditions.

IV. OUR SUGGESTIONS

Before turning to specifics, it is fair that you ask, How certain are we of the preceding analysis? Our confidence is grounded in several bodies of literature in addition to high reliability theory, which lead to our conclusions.

One body of literature is about what happens when managers and owners become distinct groups in a corporation [for instance, Gourevitch and Shinn 2005], and we have seen similar problems arising in the control room and between the control room and wraparound. Who, for example, should be responsible for using a major piece of software, such as the newest version of the state estimator—the operators or the engineers? We also know from the literature on the life cycle of the firm that units in a company can mature at different rates [see Harrison and Shirom 1999], i.e., some may continue to be in startup mode, while others have already advanced into formalizing more and more procedures. Again, we are observing this between units in the control room and the wraparound. "We're still having growing pains," said one operations engineer. "You know the *Star Trek* model of the *Enterprise,* where one guy runs the engines, one guy mans the guns, one guy navigates, and one guy's in charge of it all? I don't see that model here."

The difference in perceptions about operations among core groups within CAISO can most usefully be highlighted by the Figure 1 framework, which was based on our earlier interviews with your operators and support staff. In the latest round of interviews, it is clear to us that some staff outside the control room believe operators are working under just-in-time conditions with many options, while those in the control room are actually operating with far fewer and more volatile options, just for now. We suspect other bodies of literature (for example, human factors analysis for measuring task load and reaction time capacities under varying degrees of stress) would confirm what we are observing. These differing perceptions among key participants arising from various directions is especially troubling, as high reliability requires these participants to be on-point with respect to their message all the time. Prolonged just-for-now performance fractures that focus and undermines the required situational awareness.

Our suggestions below are grounded in our conclusion that CAISO is moving closer to institutionalizing control room operations within the just-for-now performance mode. Our suggestions are directed toward promoting and enhancing effective management within the reliability bandwidths, where it is absent, undermined, or in need of reinforcing.

Just what management within bandwidths are we referring to? Certainly more than a closer management to stay within the reliability standards for CPS and path violations. For example, what protections to the propagation of error are there in new software that promotes faster transaction speeds? What are the limits to procedural revisions in terms of operator attention and memory? As experience with managing within newly agreed-upon bandwidths improves, more accurate patterns and scenarios emerge for high reliability management.

We believe the following suggestions are the best way to move reliability into a more studied condition at CAISO.

Increased authority in the control room over introduction of changes, such as procedures and software. In other high reliability organizations, control operators are responsible not only for the safety of systems, but for the integrity of their own management and control processes. This is part of the ownership and responsibility culture we have observed in high reliability organizations. This culture includes operators having veto power over the introduction of new procedures and software when operators believe these changes threaten to degrade their reliability envelope.

Instead of passively receiving and accepting new procedures or reluctantly consenting to them, shift managers and their crews should be the first line of defense in assessing the utility of these introductions on reliability. The role of the shift manager remains central. "There is always an answer, ultimately you can shed load," reaffirmed a gen dispatcher. "That decision is kind of above my level to call; if a shift manager is in the room it is his decision." So too in the case of new procedures. Acceptance of a procedure, software, or workaround should mean that it has been actively assessed and approved by shift managers on behalf of their crews. This has implications for the next suggestion.

Increased control room ownership of wraparound services. It is clear that control room operators must be more involved in defining the needs that drive wraparound research and development. At PJM [the major independent system operator in Pennsylvania, New Jersey, and Maryland], we understand an 80-20 rule operates within their wraparound for reserving at least 20 percent of product design to operator participation. One operations engineer uses something like a 90-10 rule, with the 10 percent being changed or corrected through interaction with the control operators. We, however, observed wide variations in the ratio among different members of that department.

This involvement will put a burden of more time on control room personnel. This may add support for a sixth crew, one whose members can fill in or replace operators in other crews who are "on assignment" with wraparound colleagues in research and development tasks or attending the increasing number of meetings over procedures and reliability protocols.

If a new crew is not possible, then consideration should be given to creating a permanent unit of control room operators and key wraparound staff, headed by control room personnel, to move beyond firefighting toward rectifying current quick fixes in control room software. In some high reliability organizations, "white-hat" teams are used to develop scenarios to crash systems as a way of sharpening what precursors must be avoided [for example, Lawrence Livermore National Laboratory 1998].

If such a unit is not possible, then teams of control room and wraparound staff should be created around each major new software, hardware, or procedure, each product-team having its eye on high customer service, where the customer in this case is the control room operator. One goal of the new product-teams should be to offer full-service advice to operators and to organize the corrections and improvements needed in the control room and elsewhere (e.g., in settlements) because of the new software, hardware, or procedure.

If product-teams and processes are not adopted, an alternative would be to designate someone from the wraparound to be delegated the authority to undertake product trouble-shooting that could cross departmental jurisdictions in ways that are increasingly difficult given the specialization of wraparound departments.

Whatever the institutional format for this improved control room–wraparound collaboration, a very clear priority is the roll-out and operation of the new state estimator. From our high reliability perspective, anything that helps generate shared pattern recognition and scenario formulation for both the control room operators and wraparound staff is to be encouraged, since it builds up the domain of your reliability professionals. In the words of the engineer in charge of the state estimator imitative: "As engineers and operators use [the state estimator] more, they'll see what is the health of the system. . . . In the beginning, it will be used to identify patterns that serve as alerts and warnings and over a year or two, then it can be used for scenarios into the future." To do so, however, may well be a challenge. One senior shift manager called the new state estimator "another rock to hold" for the control room.

Institutionalize the reliability test in the wraparound. Our earlier research noted that CAISO regulators and outside regulatory agencies frequently recommended "improvements" that did not pass a "reliability matters" test: Does the proposed improvement increase options, reduce volatility, or enhance the operator maneuverability to balance load and generation as performance conditions change? Or to translate these test questions into admonitions: do not introduce volatility into control room tools without clear and direct fallback options; do not introduce new tools that diminish the ability of control operators to follow in real time what is happening; and do not introduce tools that de-skill operators who have to take over if and when the tools fail. We now wonder if some of CAISO's own proposed improvements would pass that test.

We expect, from the experience of other high reliability organizations, that CAISO's control room operators would be given the resources and tools to protect themselves from being pushed into making forced errors. Indeed we would expect the primary responsibility of the wraparound was to prevent forced operator error from occurring. To that end, the wraparound needs to operationalize a reliability test in the design of its own deliverables to the control room.

It is important to recognize that reliability cannot be a design-driven operation (the design node in Figure 2). Design has to be translated into what works for the control room operators' grid management, as your senior market operations official put it: "We give them [the operators] knowledge by translating the complexity so that they can understand it in the display. That is a challenge for everybody [in the wraparound] and the better we do it the more robust the system becomes." Another engineer put it this way: "If the wraparound takes action to clarify things, automates some of these things, boils down analysis, then this enables the operator to worry about other things." If operational skills are to remain at the center of CAISO's reliability mission, this capacity to "translate and boil down" requires the development of a shared model among all wraparound designers of the reliability skills and cognitive capacities of your control room personnel. This model is the basis of the next proposal.

Develop a shared model of control room operations and reliability skills. We noted earlier that second-guessing of operator decisions is on the rise. Clearly, people say they know the control room and the positions there, but we did not find a common understanding of the cognitive skills of and pressures on control room operators in real time. Indeed, as mentioned earlier, we are finding different perceptions among wraparound personnel of these skills and pressures.

Moreover, a refrain we have heard is that so much has changed in the control room that the current operations of the various desks would be unfamiliar to many, if not most, of their former control room operators. Not only the tasks, but the skills needed for any given position have been changing: "Now, it's a lot more personal skills, multitasking has always been there, but before conflict resolution was low and now it's really important," reported a shift manager. Thus, without a shared conception of control room operations, it is not possible to develop or implement interventions that take into account and are consistent with such changes. Since some second-guessing by wraparound staff of control room decisions is inevitable, it is important it be based on shared conceptions of operator requirements for scenarios and patterns. As a senior shift manager told us, "With the generation and transmission desks, there are a lot of things open for interpretation, whole new scenarios need to be explained."

Clearly, the more time wraparound staff spend in the control room observing the positions for which they provide support and the more time they actually spend in either actual or simulated control room performance conditions, the more realistic and shared will be their conception of these positions and conditions. This training should extend well beyond a few hours. In developing the shared model, we recommend what we observed in other high reliability organizations and have previously commended to CAISO: Extensive cross-training coupled with rotational assignments over some months across *all* departments.

Improve training curricula and modules. The need for a shared model as well as the need to reinforce the operator skill base in inexperienced recruits to the company can both be served by improvements in training. One control room dispatcher told us, "Some operators in here have the whole system in their heads. It's amazing. If you locked them in a room and gave them paper and a pencil they could sketch out the whole grid." The same expertise in cross-scale and operations analysis should be the case for those in the wraparound. And vice versa. When a senior market operations official told us that "they have all the scenarios in their heads to explain why it's still congested," he was talking about his engineers, not those in the control room. In the absence of cross-training, a shared model, and common knowledge bases, you can expect more problems like the one raised by a senior shift manager: "We're becoming all experts; so we become a single point of failure. [One key wraparound staff person] is off [on vacation]; when he goes so goes your expert. . . . If we had a catastrophic failure in EMS, it would take a lot to recover." Again: more just-for-now performance.

We participated during the 2004 summer training in a simulation exercise, which while only a small part of that week, could be an important mechanism for training to enhance scenario development and pattern recognition among operators and wraparound staff. Indeed, we believe CAISO's simulator to be a capstone resource, e.g., when the sixth crew is not working with the wraparound, it should be in the simulator. Wraparound staff should rotate through simulator training. Moreover, this training itself would be enhanced by the development of a "case history" initiative for more trainable events.

Raw material for the writing of these case histories should begin with the practice of "hot wash-ups" or post-mortems just after key failures and close calls. Many high reliability organizations formalize this material into written histories which are widely shared within the organization and which are part of their training curricula. This is particularly important for those inexperienced control room operators who come to the company with no specific control area experience. Further, the case histories are frequently shared with other comparable organizations in order to widen the scenario and pattern base available to operational personnel. This curriculum-based training can support real-time reliability by enhancing the experience base of operators. The summer training we observed seemed to be based less on increasing this experience base than on the premise that better communication skills made for a less stressed personality, which in turn makes for a more reliable professional.

We believe that the five suggestions proposed above—increased authority to control rooms, a more realistic and shared understanding of control room operations and skill sets, bringing the reliability test more fully into the wraparound, increased ownership by control room operators of wraparound services, and more focused training—will improve grid and service reliability. The urgency is to move out of just-for-now performance conditions as soon as one can, given the future for CAISO is as full of flux and change as the present and past have been.

CAISO'S THEN COO acted on our report immediately. He initiated a series of discussions in January and February 2005 to follow up its major points and suggestions. Our report, along with that of two consulting engineers (which added support to our findings), and the discussions that ensued led to revisions in how procedures were developed and approved, increased cross-training, and greater recognition that wraparound staff needed to return to their earlier support role for the control room. Difficulties continued to be encountered, but 2005 turned out to be the best performing year in the history of CAISO as measured by CPS2 (no fined violations in 2005 for the first year ever) and OTC "path" violations (similarly no violations for the first year since CAISO started operations). The multiple reasons for this are discussed more fully in the next chapter.

We would like to believe that the readiness of CAISO officials to consider and respond to our report was due to the special merger of theory and practice that we, as researchers, had demonstrated over the years to key management and control room staff. That said, it is obvious to us that the receptiveness of CAISO executives was clearly a function of our being able to explain what we were seeing in our second stage of research in terms of the framework developed with them during the first stage. As Lasswell (1951, 13) put it years before, "It is enormously fruitful for the academic specialist to take some of these ideas [of active practitioners and participants] and give them the necessary systemization and evaluation." We also believe Lasswell would not have been surprised when officials have *as a management strategy* taking advantage of independent researchers in reliability management.

After the report and discussions, we resumed our research on indicators of the reliability envelope. But things never stay the same at CAISO. "It's still continuous change," a shift supervisor told us in 2004. "It's not like building a brick wall, where it all stays the same," said a gen dispatcher, "A lot of things change here." "Things are changing, always changing," said another gen dispatcher. "It's neverending change here," reported a control room support staffer in late 2005. "[There's] always big challenges as operations evolve, always issues to be dealt with after the fact."

In 2005, a new CEO came on board, and CAISO underwent a major reorganization, with the departure of over one hundred staff including those from the control room. Our development of indicators took on even more urgency for us in 2006.

CHAPTER 6

CAISO IN 2004–2006
New Challenges and Responses

WHERE IS THE CAISO CONTROL ROOM TODAY?
The BEEPer has long gone with the introduction of
RTMA. The shift manager has become the shift supervisor with added duties,
control room operations have been consolidated with market operations, and
one part of the California grid has become its own control area (the Sacramento Municipal Utility District). In addition to its new CEO (who replaced
an interim CEO, who replaced CAISO's first CEO, whom we first met in
2001), CAISO has a new board of governors. Outside of CAISO, Enron has
long gone bankrupt, lawbreakers have been convicted, and other grid participants have "reorganized." PG&E has gone into and out of bankruptcy, while
SCE has come back from the precipice of financial insolvency. The state's
regulatory environment, although calmer, is still crowded. The August 2003
blackout rattled more than the Northeast, eventually leading to a new organizational structure for electricity reliability in the United States. In 2001, we

and our colleagues identified design incompleteness as a permanent feature of California's restructured electricity network, and we have seen no reason to change that conclusion.[1] That said, significant improvement in the stability of this incompletely designed network can be observed.

What form does CAISO's high reliability management now take? Where does the control room stand with respect to its reliability envelope, wherein skills and tasks are matched?

A number of recommendations were made to CAISO in our January 2005 report and in an emergency preparedness report delivered later to CAISO by Nexant, Inc.,[2] a consulting organization. Among these were that

- It be made clear that the shift manager (now supervisor) and all real-time operators are the primary customer of operations and market support groups in the wraparound.

- The authority that the Grid Operations unit has in adopting new procedures and tools be clarified to include a process that solicits input from all shifts regarding these changes.

- Through whatever means necessary, increased time for training for the shift personnel be provided.

- A transmission emergency exercise be undertaken to demonstrate preparedness for a transmission emergency. This would be most effective with all of the distribution utilities and key generators participating in a major simulated transmission emergency.

- A high-level support person be assigned with significant grid operations experience to be available on the floor to assist real-time operations.

These recommendations have been adopted along with others. The Grid Operations and Marketing departments have merged, control room personnel are now expected to participate in new software development undertaken in the Engineering department, and more intensive cross-training among control room personnel is being conducted so that each crew member can eventually take over any other control room job.

As part of an overall realignment, in late 2005 CAISO reduced its personnel by nearly 20 percent. This was accompanied by significant turnover in control room personnel and a major loss of personnel in the Engineering department.

One CAISO official described this restructuring as "the most significant changes since startup." Control room morale during this period, as revealed in our ongoing interviews, was low.[3]

In early 2006, CAISO announced a major new business plan. The plan was accompanied by a rollback of CAISO fees to market participants to 1998 levels. The plan conceded that "operations [have] been in reactive mode since the energy crisis" with a "focus on 'fire-fighting,' just 'keeping the lights on' and balancing the system." It went on to call for making CAISO more proactive in improving grid management practices and technology and for creating a "world class Operations Division."

New reliability goals were asserted, including a push to reduce to zero the number of DCS and OTC violations. The posting of market settlement statements to market participants (SCs) was to be reduced to ten days or fewer rather than the average lag of fifty-one days. Increased reliance on technology, such as the state estimator (a monitoring and analytic system for arriving rapidly at grid optimizing solutions) and a Market Redesign and Technology Upgrade (MRTU) to further automate grid management, was a major element in the new plan.

One intriguing piece of the new plan as explained to operators in briefings we attended was a shift in the reliability role of CAISO. Instead of searching out megawatts, sometimes at the last minute and at spot market prices, so as not to undertake mandatory load shedding, CAISO would no longer be the reliability guarantor "of last resort." What this means for actual control room decisions is as yet unclear to control operators and their supervisors.

In addition to the internal changes, major regulatory changes have been made outside of CAISO that have significantly stabilized CAISO's operating environment. In October 2005 the California Public Utilities Commission (CPUC) adopted a "Resource Adequacy" regulatory rule requiring distribution utilities in California to acquire 115 to 117 percent of generating capacity over forecast peak demand. That November the Federal Energy Regulatory Commission (FERC) approved a change in the CAISO tariff requiring all load-serving entities such as distribution utilities to schedule at least 95 percent of their forecasted load in the day-ahead market rather than contracting for load in the hour-ahead or spot markets. These regulatory changes reduce the real-time burden on control operators to scramble for added generation to cover peak-load periods. Resource Adequacy as well as several other factors have had a profound stabilizing effect on network relations.

But have these changes made a *noticeable* difference in high reliability management? We have seen that 2005 was a good performance year in terms of the official statistics, but what is happening cognitively in the control room? Are things getting better for the reliability professionals there?

IN THE SUMMER OF 2006, we undertook a detailed analysis of a set of indicators of challenges to high reliability management, factors that we had culled from our CAISO interviews since 2004. As discussed in the preceding chapter, our search was for quantitative measures of when operators were at the edge of their reliability envelope, if not over, when it comes to matching their skills with the task requirements they faced. The need for indicators was obvious. In a 2006 meeting with CAISO shift supervisors, a gen dispatcher asked if there was any way to measure the huge baseline complexity operators face and the stress it causes: "this complexity in procedures, this complexity in everything." "I dread when a major problem on Path 15 requires us to read a seventy-to-eighty-paged procedure," said another at the meeting.

We were able to gather data on those indicators for the period from July 2004 to May 2006, roughly seven hundred days. The analysis reveals just how well the control operators were able to cope with daily variations in grid conditions and buffer their impact on key reliability factors. Only during one period, from November 11 to December 31 of 2004, when the RTMA software was being introduced, did operators cross an edge and lose some of their ability to insulate their reliability performance (in CPS2 violations) from fluctuations in unscheduled generation outages, the need to mitigate (that is, reroute power across transmission lines) for line congestion, and having to bias the RTMA software to compensate for errors in forecasted load. (Chapter 11 provides a full discussion of the indicators analysis and its findings in relation to key concepts developed in Parts One and Two.)

The indicators analysis confirms how well reliability management has worked at CAISO. But it also shows how unusual conditions or significant changes in technology or software can, over an extended time, undermine the balance between the reliability skills of operators and the specific challenges they face in managing the grid.

Paradoxically, these unsettling conditions can arise from efforts to redesign technologies or regulations that are undertaken to make things better. The signing of long-term contracts with electricity suppliers by former California Governor Gray Davis is an example. Negotiated to end California's electricity crisis by providing more power to the grid and avoiding enormously expensive spot-market power purchases, these contracts posed their own special challenges to grid reliability. The contracts were signed with power producers located in Nevada, Arizona, and Mexico, far away from major load centers such as Los Angeles and San Francisco. Routing power from the remote locations to state load centers proved to be a major challenge. Operators had to find feeder lines that could handle the flow and connect them in often circuitous ways to avoid overloading the major state transmission lines. These mitigation "solutions" (at times as many as six in effect simultaneously) are complex and potentially unstable should outages or sharp load reductions occur. They tax the ingenuity and attention of generation dispatchers.

As it turned out, the contracts locked in prices that ultimately exceeded the average day-ahead cost of power after the electricity crisis, so the State of California was left with long-term commitments to pay for power at higher than desirable rates. The contracts thus became problematic both from an economic and reliability perspective. We have already noted that they did not help to save the governorship of Gray Davis, as he was removed from office by a voter referendum.

The enlargement of "reliability-must-run" (RMR) contract requirements for specific generators led to a problem of overgeneration for control operators. Because contracted power must be paid for whether it is used or not, dispatchers found themselves trying to call on contracted power whenever possible even though it might be less favorable in amount or location than day-ahead or even hour-ahead generation. RMR units pose their own requirements for special and complex grid solutions.

Last but not least are the problems posed by the spread of those Remedial Action Schemes (RASs) discussed in preceding chapters. Designed to contain voltage or frequency disturbances before they can cause cascading grid failures, RASs can themselves cause such a sudden shutdown to a portion of the transmission system that they can create their own instability in the remaining part of the grid. This in turn requires operators to scramble fast to compensate for generation or load loss in order to keep the rest of the grid operational.

ISSUES SUCH AS THESE must be part of any long-term view of CAISO's evolution as a high reliability management organization. Challenges continue and new performance edges arise to identify and avoid. Over the half decade we have observed CAISO, the organization has continually undertaken new innovations as a way of controlling its environment and stabilizing its performance. We saw that in the introduction of RTMA, and we see it in the upcoming MRTU. Such innovations initially make it difficult for operators to continue what they had normally done up to that point. There has never been a "resting place" for the reliability professionals. The challenge remains to ensure that input variance, which is large for a host of reasons, is managed so as to produce reliable outputs. We close this chronology of CAISO's development with a noteworthy example that demonstrates the considerable improvement in both reliability skills and the operational environment with respect to grid management.

We were in the CAISO control room on a hot Friday afternoon, July 21, 2006, when California set its then all-time record for electrical load: 49,036MW. The load peak was nearly 2,500MW (4.5 percent) above the previous record, set on Monday during the same week. Such a substantial increase in energy demand is an important development with far-reaching implications for California and the nation's economy, environment, and ultimate vulnerability.

An equally dramatic aspect of that load record is that it constituted a huge reliability challenge and achievement for CAISO. In contrast with high load periods during the energy crisis of 2001, this record peak was not accompanied by declared states of emergency or selective load shedding. Operators were not forced into just-this-way or just-for-now modes. In fact, control room operators fluctuated between just-in-time and just-in-case performance. They were operating well within a balance between their cognitive skills and grid management task requirements.

What accounts for the striking difference between this performance and the frenetic conditions of 2001, or even the difficult periods we observed in the fall of 2004? A major factor is clearly that the restructured electricity network surrounding and supporting the work of CAISO is much more stable than it has been. The Resource Adequacy requirements mentioned earlier (including regulated "must-run" and "must-offer" generation to CAISO) made more power available on a market basis and thus reduced the out-of-market scramble CAISO would have had to engage in to prevent its operating and spinning reserves from dropping below regulatory limits.

We also noted on this July day that informal phone communications between dispatchers and SCs at the generators about keeping units on, postponing work on units, or simply exchanging information with CAISO about the condition of generation units was much in evidence. A phone call by the VP-Operations to the BPA also revealed a willingness to send more power through the intertie if needed.[4] This communication pattern, while parallel to what we observed in 2001, was in contrast to the limited interpersonal communication we had observed more recently with the mostly formal machine-dispatch instructions from RTMA.

Possibly the increased informal communication reflected some improvement in mutual trust among network participants as well as a shared concern for the reliability challenge under unprecedented load conditions (see also Chapter 9 on some virtues of control area interdependencies). One antagonistic response, a reflection of past problems,[5] did occur when a generator responded to a request for additional power by seeming to suggest it might be available if CAISO were to pay higher rates than originally scheduled.

As load climbed throughout that July afternoon operators kept a close eye on current generation, both within and outside of the CAISO control area. One factor in grid conditions that provided some slack regarding load and preserved some options for operators was that outages due to generator maintenance were 800MW lower than forecast for the day. Further, despite some emergency conditions declared in the Idaho control area (which turned out to be precautionary only), intertie power imports were stable and sufficient.

The conditions allowed control operators to essentially max out the market (that is, use all the marketed generation), without having to cut significantly into reserves. Finally, at one point, RTMA had nothing more to allocate, thus sending CAISO looking out of market for energy. This was really a test of the peak capacity of the system. Restrictions on maintenance had kept most generators running for the entire week, and as one control room operator said, "They've been really humming away." It was noted that this was eventually going to take its toll on a lot of equipment, be it for generation, transmission, or distribution.

As load climbed to new records, we witnessed a dynamic balance between peak-load and peak-reliability performance. An intense, but confident, watchfulness prevailed in the control room. Even though load put operators in an unprecedented situation, the relatively stable conditions with respect to control instruments and options allowed operators to avoid highly reactive just-for-now operations. In fact, there was as much forward-looking, just-in-case performance as we have ever seen in our control room observations. When word of thunderstorms in the vicinity of one substation was received, immediately several control room officials, including the shift supervisor, began to formulate scenarios through the "what if . . . " format. Overall, we have to conclude that, were it not for CAISO reliability management, system volatility for the rest of the restructured electricity network would have been substantially increased that day.

The stable grid conditions also allowed operators to look ahead to scheduled changes in power generation and load later in the afternoon and to formulate strategies to counter even small amounts of generation loss. Obviously, under conditions of peak challenge and performance, even small (e.g. 35MW) changes can have magnified impact. But operators discussed these in what we took to be a patient manner because they knew they had multiple options at this scale. In fact they observed the approach of the new record eagerly and cheered when load topped 49,000MW for the first time. What a difference compared to what we observed in that same control room during mid-2001!

The long-serving VP-Ops watched this operator performance with great pleasure. After previous control room situations he had observed during which, in his view, operators were without energy or demoralized, he was enthusiastic about their confidence and relaxed intensity. "This shows the things we've been working on inside and out have really paid off," he said. Later he put it, "You guys talk about options—that day we had all kinds of options!" This frame of mind contrasted starkly with his loss for words in describing the March 8, 2004, event, which initiated our second phase of research.

Interestingly, after the peak load had passed, the grid began to lose some stability, with a line problem and a generator trip. Performance became more just-in-time, in which keeping the balance between load and generation required more real-time ingenuity. Operators actually began to cut into their operating reserves, although the spinning reserve remained above levels requiring a "warning" condition. With the VP-Ops in the control room, it seemed that operators understood that drifting a bit below 5 percent in total reserves would not force them to declare a Stage 1 emergency.

Later, in discussing the drop in operating reserves below officially allowable levels during this immediate off-peak period, several operators noted that "spinning reserve is more important anyway" and "there was really reserve out there that wasn't part of the official count. We could have found it if we'd needed it." Even in this case, official reliability measures continue to be redefined or pushed to the limits.

From the perspective of our framework, we witnessed on July 21 a peak condition in high reliability management in which unprecedented grid conditions were managed through performance that remained well within the cognitive balance between operator skills and task requirements. Although load might have been in uncharted territory, the operators' skills and options were not. Thus, despite some heightened risk in grid conditions, reliability management itself was not at risk.

That operators avoided this risk is in sharp contrast to the conditions we observed in the fall of 2004. In this respect, CAISO's handling of the load records of July 2006 stands as a remarkable organizational and managerial achievement.

PART II

HIGH RELIABILITY MANAGEMENT
Key Concepts, Topics, and Issues

THE FRAMEWORK FOR HIGH RELIABILITY management developed in Part One is focused on reliability professionals and the performance modes they are expected to operate within and across. From this framework comes a set of key concepts, topics, and issues for high reliability management, which we develop here in more detail:

- The critical balance between trial-and-error learning and failure-free performance

- Strategies of managing performance within controlled margins (bandwidths) as opposed to strategies for fixed and unwavering performance

- The special domain of operational risks as opposed to analyzed risk, a domain in which risk seeking can actually enhance reliability

- The paradox concerning anticipation, resilience, and robustness, or, the illusion of "readiness"

- The special threats and challenges to high reliability management of dominant strategies of design and technology

- The pluses—not just the minuses—of real-time operating conditions under just-in-time performance, and the potential reliability benefits of cost-effectiveness considerations

These topics are grouped and discussed in the following four chapters, which draw from and extend the CAISO material. Chapter 11 concludes with a discussion of the findings from our indicators analysis relevant for the concepts discussed in Parts One and Two.

CHAPTER 7

ERROR, RISK,
AND BANDWIDTH
MANAGEMENT

R EADERS OF MANAGEMENT LITERATURE HAVE
been told for years that if complex organizations are to
survive, they must learn by embracing error. Don Michael, Warren Bennis,
Edgar Schein, David Korten, and Robert Chambers all have argued that an organization's capacity to persist and grow depends on its ability to learn from
mistakes and failure. As Michael put it in the 1970s, "Future-responsive societal learning makes it necessary for individuals and organizations to embrace
error" (1973). In the early 1980s, Korten maintained that the best organizations
were ones "with a well developed capacity for responsive and anticipatory
adaptation—organizations that: (a) embrace error; (b) plan with the people;
and (c) link knowledge building with action" (1980).

As early as the 1980s, Bennis concluded from his studies on leaders that
they must "embrace error" and not be afraid to make mistakes, admitting

them when they do (Bennis and Nanus 1997). Schein sums up, "We come to embrace errors rather than avoid them because they enable us to learn" (1994). More recently a *Harvard Business Review* article, "The Failure-Tolerant Leader," reported that such leaders "don't just accept failure; they encourage it" (Farson and Keyes 2002, 66).

Nothing in this is new. As Frederickson and LaPorte (2002, 33–34) remind us, error tolerance has long been a hallmark of organization theory:

Much of what we know about public organizations and their management is based on the study of error-tolerant organizations, with their difficult-to-measure social purposes, ambiguous political messages, limited resources, trials and errors, and therefore, tolerance for failure. . . . Based on the study of error-tolerant organizations, theories of buffered or limited rationality, muddling through, mixed scanning, incremental decision making, contingency, efficiency and sense-making have dominated our literature.

One major problem, of course, is that "error tolerance" is often involuntary. Embracing error looks quite different from the perspective of those already caught deep in its clutches. That said, the theme of learning, be it through unintended trial and error or by virtue of the deliberate "Experimenting Society" (Campbell 1988), continues to have special resonance and purchase with social scientists and organization theorists.

Yet Frederickson and LaPorte (2002, 35–36) describe a special class of organizations that are better thought to be error intolerant. These are what they and we have been calling high reliability organizations (HROs):

The perspective of highly reliable organizations is based on many years of direct observation of error intolerant systems, such as air traffic control, nuclear power generation, nuclear submarine and aircraft carrier operations, production of the components of nuclear weapons, and electricity transmission systems. . . . By any measure the safety and reliability of these systems is remarkable. . . . This unique class of organizational systems works in the context of the essential insistence that they be nearly failure free, and with rare exceptions, they are.

For HROs, the admonition "Encourage error!" is a non sequitur of the first order. For them, the first error could be the last trial (Rochlin 1993, 16). The high reliability managers at CAISO would also have great difficulty with classic trial-and-error learning applied to a control room.

We have seen that high reliability organizations are characterized by the "inability or unwillingness to test the boundaries of reliability (which means that trial-and-error learning modes become secondary and contingent, rather than primary)" (Rochlin 1993, 23). Although HROs have search-and-discovery processes, and quite elaborate ones, they do not undertake learning and experimentation that expose them to greater hazards than they already face. They learn by managing within limits or by setting new limits in which learning can take place (T. R. LaPorte, personal communication, 2002). As Rochlin puts it, high reliability organizations "set goals beyond the boundaries of present performance, while seeking actively to avoid testing the boundaries of error" (Rochlin 1993, 14). Trial-and-error learning does occur, but this is done outside primary operations, through advanced modeling, simulations, and other ways that avoid testing the boundary between system continuance and collapse.

At first glance then, HROs look little like the learning organizations touted by Senge (1990) and others. The difference can be best described this way: from an error-intolerant perspective, an HRO is only as reliable as its first failure; from an error-tolerant perspective, the learning organization fails every time until the last one. And yet there is something misguided about a hard-and-fast distinction between error-tolerant and error-intolerant organizations. It is not just that learning is going on in both error-intolerant and error-tolerant organizations. Two other qualifications hold.

First, to describe an HRO requires one to describe some of the very elements observed in error-tolerant organizations. For example, the authors of "The Failure-Tolerant Leader" (Farson and Keyes 2002, 66, 69) found in their research that such leaders "openly admit their own mistakes rather than covering them up or shifting the blame. . . . Similarly, managers [at one company studied] routinely reinforce the company's mistake-tolerant atmosphere by freely admitting their own goofs."

The same can be said, almost word-for-word, for HROs. LaPorte, in his review of research on high reliability organizations, found that

HROs exhibit a quite unusual willingness to *reward discovery and reporting of error,* without at the same time pre-emptorially assigning blame. . . . This obtains even for the reporting of *one's own error* in operations. . . . The premise is that it is better and commendable for one to report an error immediately than to ignore or to cover it up (LaPorte 1996, 64; italics his).

The ready admission of error and mistakes has been found during the reliable provision of services in earlier reliability research on water supplies (Van
Eeten and Roe 2002). Whether in an error-tolerant or error-intolerant organization, the admission of error is promoted as a strategy to reduce the probability or magnitude of failure.

A second qualification also blurs a hard-and-fast distinction between error
tolerance and error intolerance in organizations. Much of the overseas development literature, for instance, is about rural peoples and governments learning
to provide critical services reliably. We find many instances in which the vari-
ously termed "learning process approach," "process management," and "action-
research methodologies" are recommended for the planning, design, implementation, and operation of water supplies, health systems, government budgeting,
and financial services, among other critical services (Roe 1999). Given the long-
demonstrated problems with "blueprint development" and rigid project designs,
and the better empirical record with respect to adaptable and flexible approaches to project development, it would seem that error tolerance must be
vital to eventual error intolerance, that is, "learning to avoid failure."

Whenever one trips over a paradox, the situation cries out for specific
cases. Our longitudinal study of CAISO allows us to take up the distinctions
between error tolerance and error intolerance and shows how both work together in the case of one extremely important critical infrastructure. As we see,
learning to set limits and manage within limits, to use LaPorte's earlier phras-
ing, is the crucial precondition for those operators and managers whose pro-
fession it is to balance error-tolerant and error-intolerant modes of operations
in achieving high reliability.

EARLIER HRO RESEARCH highlights cases in electricity transmission
and generation under the integrated utility format. For LaPorte and Lascher
(1988, 4), Pacific Gas & Electric (PG&E) was "a prime example of a class of
'high reliability' organizations that take on an extraordinary goal—to operate
complex, often hazardous technical systems in a nearly failure-free manner. . . .
A characteristic feature of these organizations is the relatively clear specification of operational failures and the consistent efforts to reduce their number
to very low levels." To that end, PG&E's control room operators managed the
flow of electricity within strict limits:

Reliability is a key operating objective of [PG&E], with particular emphasis on assuring adequate power supply in all kinds of conditions. . . . The margins for error are small. The goal is a constant level of voltage with margins of plus or minus (+/−) 5 percent of nominal voltage. Wider fluctuations could result in very costly failures and consumer and regulatory difficulties (LaPorte and Lascher 1988, 6).

Schulman (1993a, 369) made a similar observation in his case study of Pacific Gas & Electric's Diablo Canyon nuclear reactor:

The proposition that emerges from analyzing Diablo Canyon is that reliability is not the outcome of organizational invariance, but, quite the contrary, results from a continuous *management of fluctuations* both in job performance and in overall departmental interaction. It is the containment of these fluctuations, rather than their elimination, that promotes overall reliability at Diablo Canyon. . . .

It is precisely this management of electricity fluctuations within limits that has been highlighted in our case of high reliability management after the California electricity restructuring. We have described, for example, how the CAISO gen dispatcher manages the grid in real time by estimating how much of an increase or decrease in energy is needed to control the Area Control Error (note that term!), which shows the relative balance between generation and load on the grid.

Because balancing load and generation is a real-time team activity, many errors and mistakes made with respect to that balancing exercise become visible in real time as well. From the gen dispatcher's perspective, when the ACE drops 4,000MW, it has to be a scheduling input error (that is, no generator suddenly going offline could account for such a huge loss); when it drops 600MW, it has to be a generator, and very likely one from among a specific set of generators.

The timely admission of control room misjudgments is essential, and we have heard such admissions throughout our research. We observed a real-time market operator making a mistake that was obvious to those around her, and she readily admitted it. In another instance, a generation dispatcher replied to our question about why the ACE was continuing into a violation phase with "I miscalculated." When a shift manager asked how things were going, another gen dispatcher admitted to having erred earlier in the morning, which caused overgeneration with which they now had to cope. Neither gen dispatcher felt compelled to argue the mistake.

At a CAISO conference meeting in 2001, a very senior official reiterated that "in our [CAISO] culture, if a number is bad, we try to find out why, instead of blaming each other as [some state government officials] do." The ability to identify big mistakes before they happen and admit error after mistakes reflects an important fact of life for reliability professionals. They have to work with each other to reduce the big risks. Error intolerance requires error tolerance; they have to be balanced in order to reduce risks.

RISK IS A WORD many readers think they already know. Our framework gives it a more complex meaning, which in turn throws light on error tolerance and error intolerance.

The formal definition of risk is the product of the magnitude of a hazard times the hazard's probability of occurrence. Another, less formal definition of risk is the chance of something happening that has never happened before. Both these notions of risk are only the starting point for reliability professionals.

To appreciate this, start with the common perception of technological risk as the risk inherent "in" a technology. As we have seen, a fundamental premise of normal accident theory is that large technical systems, which are tightly coupled and complexly interactive, are risky if only because low-probability events can amplify into highly consequential disasters. But this risk is really a managerial one primarily associated with what we have been calling just-for-now performance when few if any options exist for protecting against failure cascades. Just-for-now, however, is not the only performance mode. Our research suggests that *diverse* risks attend the operation of the *same* technology and that these risks vary across performance modes. Although deviance amplification is the major risk in just-for-now performance, it is not the major risk in the other three (Figure 7.1).

Repeatedly, we were told in our interviews that the major risk in just-in-case performance is that the operator grows complacent and is not paying attention, when he or she should be, to unexpected changes in system volatility or options availability. The big risk under just-in-time performance conditions is the error of misjudgment under the pressures of time with "too many balls in the air," as the interviewees put it. The major risk in just-for-now condi-

System Volatility

		High	*Low*
Network Option Variety	*High*	Risk of misjudgment with too many variables at play	Risk of inattention and complacency
	Low	Risk of exhaustion of options, lack of maneuverability, and cascading error	Risk of failure in complying with command-and-control requirements

FIGURE 7.1. Major risks by performance conditions in control room (from Table 3.1)

tions is, as we have seen, that there is no more room to maneuver; options become exhausted and the system drifts into failure when load and generation are no longer in balance. On the other hand, the big risk in just-this-way performance is not gaining control and compliance over what must be controlled for reliability.

Another way to think about the differences between these risks is to recognize that each performance mode has its own precursor zone in which reliability skills and their match to tasks become endangered. Not only do managers need to attend to external conditions, they must also be sensitive to the possibility that what has worked well in that performance mode might be carried too far. In this way, the precursor zone means that

- What is routine and procedural during just-in-case performance now becomes complacency[1]

- What is adaptive equifinality and creativity during just-in-time performance becomes misjudgment and overconfidence

- What is meant to be only temporary and expedient during just-for-now performance becomes a prolonged loss in maneuverability

- What is compliance during just-this-way performance becomes rigidity and inflexibility

Rather than one major risk, what is risky varies considerably for reliability professionals. It is not that the technology makes the risk as much as risk emerges out of and varies with the performance conditions for technology's operation. The CAISO control room is one of those organizations that both manages and *produces* risk (see Hutter and Power 2005, 3). The way it manages risks adds the risk of managing that way. In this manner, high reliability management should be seen as its own unique form of risk orientation and appraisal, in which reliability-seeking behavior is not to be equated with, say, conventionalized risk aversion.

One illustration of managing risk lies with the Area Control Error. Managing the ACE within the parameters prescribed by the regulators as well as rethinking what those parameters should be in light of having reliably balanced load and generation even outside official parameters is what we have been calling bandwidth management. Note that there are *de jure* and *de facto* bandwidths at work. Some limits are given literally in the form of ACE-type bandwidths, while other parameters function as *de facto* bandwidths set outside the transmission grid, such as air and water quality standards to be maintained at certain generators during set periods of the year.

WE SEE THE MANAGEMENT of operational bandwidths for reliability purposes as taking two ideal forms. The first is founded on an effort to fully specify the system under management and to gain control over all of its critical "inputs" and thereby stabilize outputs. For each anticipated change in inputs there is a designated protocol for responding that controls potential fluctuations in output. These protocols are highly error intolerant. Even small diversions from them receive massive attention and regulatory scrutiny. In this approach, anticipation is critical, and reliability lies in a process of closing off the system as much as possible to factors that would disturb the applicability of prior analysis. Such behavior was found, for example, in the earlier research on nuclear power plant operations (such as Schulman 1993a). For ease of reference we call it bandwidth management within prior, anticipatory analysis.

A second type of bandwidth management substitutes reactive capacity for foresight. Here managers can neither predict nor control fluctuations in inputs. The system under management is too dynamic to be fully specified or controlled in all its system states. Reliability here lies in being able to undertake, if necessary, bandwidth management outside prior analysis and anticipation—that is, the ability of managers to respond in ways that buffer or tolerate input variance, including that arising from mistakes or errors, and then act to counter the variance in order to reduce output fluctuations to manageable levels. The second approach, which we term bandwidth management under active analysis, reflects the need to accommodate the unexpected and hitherto unanalyzed as opposed to management within only that which has already been analyzed.

The differences between the two bandwidth management strategies can be usefully cast in ecological terms. The degree to which an ecosystem is altered when the exogenous environment changes and the degree to which that ecosystem returns to its previous configuration once the perturbation is removed are common ways ecologists define ecosystem resistance and resilience (for example, Knapp, Matthews, and Sarnelle 2001). The resistance of the bandwidths and the resilience of adjustments best distinguish the two types of bandwidth management. In the first type, we see resistance to outside-induced shocks, that is, the system resists input variations to keep them within a prescribed bandwidth and programmed adjustments are then made to stabilize outputs. In the second type of bandwidth management, we see resilience in the adjustments, that is, managers are able to use or generate unforeseen options and resources to "rebound back" if conditions breach the bandwidths. The two types of bandwidth management, which were both found in our case study, rely on resistance and resilience differently, but are driven by the same centripetal gravity of the task at hand, namely to maintain reliability of services with the resources available and the volatilities faced in managing the large technical systems concerned. In ecology, the stability of the ecosystem depends on the amplitude of induced fluctuations, where lower amplitudes are identified with more stable systems (Ives and Carpenter 2007); so too when it comes to reliability in the bandwidth management of our large technical systems. (We return to the topic of anticipation and resilience in the next chapter.)

Why the need to have two types of bandwidth management? Because operators, as we described in Part One, continually face situations in which they cannot keep the system within the bandwidths that are handed to them. First, there are times when the larger system may itself be unpredictable, uncontrollable, or both. Rising temperatures introduce exogenous stress on hydropower systems mandated to provide water across multiple uses, including agricultural, urban, and environmental. Second, there are times when control room responses may themselves increase volatility (namely, unpredictability, uncontrollability, or both). Mitigating one transmission line may engender problems for mitigating another path, or the gen dispatcher may have biased too much in the morning with consequences throughout the afternoon. Third, there are competing reliability mandates—if you will, competing bandwidths—that control rooms are being asked to reconcile. The control room of CAISO has different mandates than those confronting the control rooms of the distribution utilities in the restructured network (such as those of PG&E or SCE). The overall set of bandwidths faced by operators across the restructured network has to be reconciled or otherwise meshed to ensure reliability of the grid.

A good example of the differences between the two types of bandwidth management comes by comparing and contrasting the bandwidth management within prior analysis found at Diablo Canyon to that of the bandwidth management under active analysis observed in the California restructured electricity system—both of which seek and achieve high reliability provision of electricity. In particular, while earlier research found that HROs avoided anything like the large-scale experimentation observed in the California ISO, the latter's improvisational and experimental operations have been crucial to meeting its reliability requirements in real time.

Again and again, we observed CAISO control room operators and their support staff trying out things that did not work before they struck on what actually worked, just in time. Such probes create space and opportunity for operators to generate new options for balancing load and generation to add to their existing repertoire of other ones. In the building up of this repertoire of "equifinal options" is where we see trial-and-error learning most evident (again, this does not mean that all options are equally effective!).

Although bandwidth management under prior analysis is intolerant of error, its major risk arises when inputs and outputs are no longer stable, namely, when conditions actually require bandwidth management under active analysis. Bandwidth management under active analysis, in turn, tolerates risks and errors we would expect to arise for an organization whose high reliability management depends on its maneuverability across performance modes as conditions change unpredictably or uncontrollably.

It is noteworthy that bandwidth management, whether within prior analysis or under active analysis, differs significantly from conventional approaches to management found in organizational theory. On one side are what could be called "all-adjustment-no-bandwidth" management approaches of the learning organizations described at the start of the chapter. You see this type of approach prominently in those advocating "adaptive management" as key to environmental restoration. Here the focus is on management as a learning process or continuous experimental process during which incorporating the results of previous actions allows managers to learn constantly for better resource management (Van Eeten and Roe 2002). Adaptive management, for example, does not assume best practices as the starting point. That is for experimentation to confirm or disconfirm. If you think of bandwidth management as "management within tolerances," adaptive management approaches may, quite literally, be outside known or negotiated tolerances.

On the other side are "zero-bandwidth-no-adjustment" management approaches found in much of the early literature on organizations. Here it is said the organization must act invariantly, in the "one best way"[2] or only this way and all the time. The organization's technology may be so hazardous that under no circumstances must the organization act as if it were a learning organization through trial and error. Everything is to be by rule, and each rule is to be based in full causal knowledge of the system. Operators have zero flexibility, zero room for discretion, in this regard. Interestingly this approach has also been taken by some regulatory agencies in their demands for organizational compliance with health, safety, or anti-trust regulation (Bardach and Kagan 1982).

In practice, the critical infrastructures we have reviewed and studied must navigate their way to reliability between those siren calls made for all-adjustment organizations and zero-bandwidth organizations. On one side, what sounds compelling starts with, How can an organization survive if it is not a learning organization? On the other side, How can the risks of error, any error, be tolerated in the tightly coupled, complexly interactive technologies critical to our society? In reality, our critical infrastructures navigate between both error tolerance and error intolerance through risk management strategies that work for them and the rest of us.

CHAPTER 8

ANTICIPATION, RESILIENCE, AND ROBUSTNESS

MAJOR DISCUSSIONS ACROSS MANY FIELDS
are under way about resilience and resilient organizations, be they critical infrastructures in the public sector or firms in the private sector (Sheffi 2005). A literature already exists on the differences between anticipation, resilience, and robustness in organizations (for a start, see Wildavsky 1988). The terms are, however, beset by the ambiguity of natural language. They have been used inconsistently and more often than not imprecisely (see the review of "resilience" in Glicken 2006). *Resilience* has been used to describe the ability of an organization to absorb shocks in order to maintain a steady state and also the ability to rebound back from shocks to a new steady state. Given that, what isn't resilience?

Understanding the cognitive domain of reliability professionals at the operational level of critical infrastructures brings clarity to the notion of resilience

and enables us to distinguish it from other popular notions, including anticipation, robustness, and even recovery. We provide that clarity within the
framework of high reliability management, presented in Part One. Our approach is to identify the properties associated with resilience, anticipation, and
robustness but from an interior view, distinguishing each by the different cognitive approaches it requires of system operators. With that foundation, we can
then analyze what larger organizational features are required to promote and enhance these operator capacities and orientations. In this way, we hope to provide
a clearer understanding of what it actually means to talk about a "resilient organization" for the case at hand. Our aim is not to provide general definitions of
organizational *resilience, anticipation, recovery,* or *robustness,* but rather to as-
sign the terms a clear meaning in at least one major context.

As we have seen, reliability professionals operate in a cognitive space driven
by pattern recognition and the formulation of scenarios for action under diverse contingencies. How is this actually done? We started our answer in Part
One, but now we need to address it more fully. Just what specifically do reliability professionals do in their domain of competence? It is one thing to recognize patterns and formulate action scenarios as useful cognitive strategies in
the promotion of reliability. But what is that "knowledge," and what are the
processes for knowledge formation in this cognitive space?

START WITH THE TERM *reliability professionals.* It is plural, and deliberately so. Professionals work together in their domain of competence. They
may see themselves as individual operators—dispatchers, schedulers, controllers,
technical support personnel—but they are networked into crews, teams, and
support staff. It is therefore a combined cognitive space that we describe. Second, in the same way the professionals are networked, so too is the knowledge
base interconnected across their individual domains of competence. Change
the network of professionals, say by changing one professional and his or her
knowledge, and you affect the interconnected knowledge base.

One gen dispatcher discussed the team nature of his knowledge base in
terms of distinct types of individuals and their learning styles within and between crews. One approach is technical. "These people want to know the ins
and outs of everything. They want to know how and why things work—the

principles behind them." They worry about a problem until they or others figure it out. Next are the "I believe" people. These are people who follow rules
faithfully and trust in previously established patterns. Then there are the "bold
types"—people willing to investigate things by trying something new. Finally, he
noted, there are the people who are cross-trained to bring more than one perspective to a problem or practice. "There should be more of these," he asserted.

These "types" constitute a team dynamic in the knowledge-building process of reliability professionals. The last two types enhance pattern recognition
by adding to the inventory of observed patterns, either by constrained trial
and error or by importation from other areas of practice. They can also add to
the scenario inventory through these processes. The technical people validate
the new knowledge and make it official. They can reinforce experiential insights through formal explanations and translate them into the language of designers and engineers. Finally, the "I believe" people are the custodians of the
old knowledge. They sustain it in memory and practice. They make sure old
ideas are not forgotten or abandoned prematurely in the process of advance-
ment and adaptation.

These types of individuals and learning styles collectively enhance one another in adding to the cumulative pattern recognition and scenario formulation among and between crews. We were repeatedly told that each crew has its
own personality, which across a full crew rotation can work to produce multiple ways to achieve the same ends. This gen dispatcher's typology neatly describes why the knowledge base is a team property and not an individual one
among reliability professionals.

It is very difficult to define formally what the shared knowledge base consists of at any point in time. One virtue of our longitudinal case study has been
the demonstration of just how dynamic, interdependent, and numerous are the
people and skills in the networked knowledge base of reliability professionals.[1]

Even with their networked knowledge base, reliability professionals sometimes encounter operating circumstances not covered by prior contingency
scenarios nor consistent with already recognized patterns. A recognized pattern can help to determine what does not fit that pattern. "There's more attention on what's abnormal," a BEEPer told us in 2004. "Operators constantly
look for the abnormal." So too do those in the wraparound assigned to give
real-time support to operators. "If I see things I didn't expect," a well-regarded

engineer told us, "something occurred and my brain says it shouldn't happen; if it's counterintuitive, then I get concerned." In such instances, there may be no prior anticipation available on which to base a specific response—no informal routine, no best practice, no formal design protocol, no emergency plan. No prior contingency scenario may have been formulated for such a situation, nor are current practices and routines necessarily helpful or even appropriate in responding. In such circumstances, reliability professionals still have to respond and adjust their behavior in order to ensure reliable service provision.

The CAISO control room, for example, may encounter a string of days of unprecedented high temperatures, then a hitherto unseen combination of generators going offline unexpectedly with a key transmission line failing, along with a formal design feature (for instance, a telemetry system with erroneous values) that actually makes things worse. In this sequence of events, the appeal to the inventory of recognized patterns and contingent scenarios fails to cover the existing situation. Nonetheless, control room professionals must respond and respond rapidly. They adjust, and these adjustments and responses, if they actually work, end up adding to their cognitive repertoire by identifying new patterns, formulating new contingency scenarios, or adjusting existing ones.

Fortunately, the skills that professionals have honed promote active analyzing and rapid adjustment in unfamiliar circumstances (Klein 1998; Weick 1995). Also, it is fortunate that a dispatcher who must make adjustments in the absence of specific patterned routines and contingent protocols is doing so within the context of cumulative knowledge among networked professionals. Other professionals in the network may know something more about what to do, even if not one of them has ever managed under this specific set of circumstances.[2]

In other words, even when reliability professionals confront circumstances in which prior recognized patterns do not cover what is happening and the inventory of contingent scenarios is insufficient as well to guide action under the observed conditions, informed adjustments and considered responses are possible. Pattern and scenario have to be mutually developed and adjusted.[3] In this way, new adjustments are developed to explain and accommodate novel conditions. This shared capacity to adjust in the face of absent or inadequate patterns and scenarios is what we mean by the *resilience* of the organization.

A LANDMARK ANALYSIS by Aaron Wildavsky (1988) offered a fundamental distinction between anticipation and resilience.[4] Following Wildavsky's
lead, many treatments of resilience have maintained this contrast between the
two properties. Anticipation, from a reliability perspective, is a design-level
strategy to forecast possible challenges or threats to the operation of a system
and to defend against them through design, plans, or procedures. A resilience
strategy, on the other hand, trades off attempts at foresight for an enhanced reactive capacity to cope with unanticipated threats once they have occurred.
The trade-off occurs because attention and resources either are frontloaded in
designs that specify threats explicitly and constrain system behavior to address
them, or are focused on maximizing flexibility to cope with a variety of threats,
some unforeseen or ambiguously defined.

But the reliability professionals we have studied, under pressure to adjust
rapidly and be resilient, are also anticipating the next step ahead. A gen dispatcher told us, "I have to be two or three decisions ahead (three on a bad day).
I have to be aware: what are my options?" Another gen dispatcher summed up,
"I have to worry about the future at the same time I'm doing the present." In
this way, resilience *includes* anticipation, but a bounded anticipation, literally
a contingency analysis of the next steps forward.

How does this work? Recall bandwidth management and our analysis of
its two types. Bandwidth management under prior analysis, which we have
called "resistance," is a special case of anticipation. Resistance means that contingencies have been planned for and controlled through a wide range of designs and strategies that ensure both low input and low output variance.
Again, the operation of a nuclear power plant is exemplary.

But there is also bandwidth management under active analysis, which is
what we are now calling resilience. Here reliability professionals, as in CAISO's
control room, make adjustments to bring the system back within the reliability bandwidths from states that are not covered under previously recognized
patterns or within the inventory of established scenarios and action protocols.
This may occur because of novel system conditions, or conditions that place
operations outside regulatory bandwidths. In this process, operators are able
to discover and create new options as needed within their task environment.
Once again we see high reliability management as its own form of risk orientation and appraisal.

THE MERGING OF ANTICIPATION AND RESILIENCE at the operational level, as found in bandwidth management under active analysis, helps us better understand important risks originating outside the control room that have direct bearing on reliability management. As one moves from the operational level to upper management and policy levels (even within CAISO), anticipation strategies become much more differentiated and distinct from resilience strategies. The two diverge, with priority given to anticipation as a design-and-control strategy. As one goes higher in the organization or outside to regulators and legislators, anticipation in the form of more comprehensive planning, law, and regulations moves center stage. Sometimes, though less frequently, these planning initiatives may include efforts to enhance the resilience of the system (for example, to increase reserves or make service-level contracting more flexible).

It has also been our observation that the higher up or farther outside you go from the control room, the more likely anticipation and resilience are to be functionally set at cross purposes. This is especially pronounced at the policy level, at which the stated aim is frequently to ensure "readiness" by maximizing both anticipation and resilience, only to find that the funds are not there to do both. Rather than anticipation and resilience merging as they do at the cognitive level of reliability professionals, system planners and designers then see a trade-off between anticipation and resilience, in which time and money spent on one necessarily means less spent on the other.

In these circumstances, arguments are frequently made at planning and policy levels that it is better to anticipate every possible contingency (or easier to sell the idea of such anticipation) than maintain rarely if ever used "excess capacity" for emergencies. This stance threatens the ability of reliability professionals to handle the risks associated with these kinds of macro-design approaches. What is "excess capacity" to the outside is bandwidth management to the inside. The divergence of anticipation and resilience as one moves up and out of the control room of an organization or network is paradoxically one reason why macro-design so frequently poses risks that have to be managed by cadres of reliability professionals, thus reinforcing all the more the need for managing within a bandwidth framework.

The merging of anticipation and resilience at the operational level also helps us better understand the nature of risks actually faced by professionals

when they work under different control room performance modes, as discussed in the preceding chapter. Return to Figure 7.1, and you can see how
these major risks arise when anticipation and resilience do not work together.
Just-this-way performance is risky because it is predicated on anticipation with
little or no resilience. If there is noncompliance, when command and control
do not work as anticipated and when no other options are available to adjust
to the unexpected, little can be done to rescue the situation. On the other
hand a major risk in just-in-time performance is too rapid a response with errors caused by too little prior anticipation.

The major risk in just-in-case performance, complacency, can be thought
of as the lapse in the focus of the professional to bring anticipation down to
the level of the next step ahead. Operators may fail to bring available options
to bear when they are actually needed. Just-for-now performance, in turn, is
vulnerable to the risks of cascading events whose actual unfolding is so rapid
and complex as to defy both anticipation or immediate adjustment.

In brief, when anticipation and resilience are isolated, differentiated, or
otherwise rendered distinct, whether outside or inside the control room,
heightened risks to reliability management should be expected.

RECOVERY OF A SYSTEM after failure may look like a special case of
resilience (Allenby and Fink 2005, 1034), but the differences are instructive.
They also are important because the system failure we are discussing—the
nearly complete collapse of a major electrical grid—is typically considered a
crisis that tests the resilience of systems to manage and recover.

Very few people among the many we have interviewed at CAISO have experienced an actual system recovery from a full grid failure, and all interviewed
about recovery have said they hope never to experience one. Thus it remains a
matter of speculation as to how reliability professionals would proceed under
such circumstances in the California grid.

The following analysis is based on our observations and interviews at the
first-ever simulation exercise for "black-start" system recovery sponsored by
CAISO in May 2006. It involved control room staff from CAISO, the three
major distribution utilities in its control area (PG&E, SCE, and SDG&E),
and control room staff from adjacent control areas, including Los Angeles. A

black-start is the reassembly of the entire electrical grid after its breakdown into an isolated set of "islands" of power generation without a usable electrical flow between them. Generators in many instances have to be restarted without any existing power flow upon which to rely. The grid has no voltage or frequency, the major indicators used to monitor its condition.

What we observed during the day-long exercise was this: if the large technical system fails entirely, its recovery by reliability professionals requires *a new performance mode* for the management system, one that comes into effect when the other four major performance modes are no longer operative. This shift has important implications for our understanding of the professionals' domain of competence, though the analysis must necessarily be treated as suggestive.

What specifically is recovery? Recovery finds reliability professionals operating under conditions that have drastically altered their cognitive space and domains of competence. The two dimensions of scope and knowledge base are radically condensed into what we would describe as a cognitive singularity. All attention in the system collapses to a single case, the case of recovery. In recovery, reliability professionals must focus on restoring grid reliability before they can resume service reliability mandates. In this manner, reliability professionals gain an unusual clarity of objectives in recovery, since they are not trying to address service and other load issues at the same time.

As such, recovery cognitively differs from resilience. Resilience is, as we have seen, the capacity to adjust or return performance in unknown conditions back to and within the existing bandwidths set for reliability. The bandwidths remain the center of gravity for resilient reliability management. Recovery starts literally when the bandwidths and the normal means of measuring and controlling where the system is in relation to them are not functioning. In such a configuration, the CAISO control variables would be inoperative in relation to the grid's condition. The task then would be to restore the grid itself back to a state in which control variables once again matter and it can once again be operated within and around its reliability bandwidths.

How did the grid recovery exercise actually work? At the start of the process there was a sharply reduced role for CAISO. Each distribution utility must restore that part of the California grid whose load it is responsible for, that is, PG&E for northern California, SCE for much of southern California,

SDG&E for San Diego, and LADWP for Los Angeles, including smaller utilities and their control areas. Only when these local areas are up and running can CAISO direct their reintegration into a grid.

Each utility in the recovery exercise we observed had its own team, and each team used its own grid outline map(s) to develop, transmission line by transmission line, breaker-by-breaker, bus-by-bus, the sequence in which its own grid would be (re)energized. Each decision—this line rather than that line—was marked in red on the gridline map as energized before the next decision was made. One or two in each team took the lead and seemed more knowledgeable, but all decisions we observed were discussed out loud and in ways such that other team members could add their input before the decision was taken and the line marked as reenergized. A utility might have its own documentation for the restoration sequence—PG&E had a handbook, SDG&E had a checklist—but these were referred to only periodically as reminders of what to do.

Reminders also came from CAISO to the utilities throughout the day, though more frequently as the day progressed and the challenge of reconnecting the separate grids into an entire California grid came closer. Members of each utility team were in regular cell phone contact with CAISO transmission liaisons assigned to them during the restoration. A great deal of what we observed has to be considered artificial—everyone was in the same room, and some critical software and cell phones were assumed to remain functioning. Restoration of the market was not part of the exercise, and as one participant noted, "If this were real we'd be under pressure early on to get the markets back up."

Activity levels were intense—even during the breaks we would see one or two team members looking at their respective gridline maps, working out the moves ahead. The challenge in each team was to find a scenario that restored their grid as fast as it could but not so fast that amplifying errors would be initiated. In the recovery process a single error in line reconnection and energizing can cause the entire subsystem to collapse back to zero in electrical flow. "We have to take it slow," "One feeder at a time, I don't want to trip anything," and "This is a house of cards" were comments we heard to that effect.

The recovery becomes itself a contingency scenario evolved in action. It does not precede action. The contingency scenario consists of the operations actually undertaken and unfolds as the professionals continue restoration. For these reasons, the contingency scenario under recovery is not a preset protocol; it is enriched by everything the network of reliability professionals participating in the exercise bring to it.

Although the entire exercise constituted a single case, fundamental macrodesign principles did not cease to be relevant. They surfaced in reminders made to the teams by their facilitators throughout the day. Indeed, reminders from CAISO to the utilities were about basic principles and design considerations (such as the use of load to stabilize frequency) that could not be taken for granted by the respective teams in emergency conditions and needed to be readily present in team decision making during this special time. In several situations, principles came into play that you do not see in normal operations (for example, the "Ferranti effect," a short-term but major increase in resistance within a previously open transmission line under new load). Basic principles also surfaced on their own in team decision making. As the teams developed step-by-step their scenario for restoring operations, members would invoke principles that applied directly to that part of the scenario they were developing—here are buses, here are breakers, here's what we have to do now.

The point, however, is that fundamental design considerations were brought to bear that are not generally part of day-to-day operations. At the same time, these basic principles were focused quite narrowly on the specifics of the situation at hand. As one shift manager at CAISO described it, "I never stared so hard at such a small piece of line." What we most certainly did not see at all during the day was a programmed course of action generated from prior principles. A cookbook orientation would be disaster, we were told by one control room participant.

Two related observations impressed us. First, no recovery scenario proceeded the same way. Different teams across the different weeks of training restored their grids in different scenarios during the simulation exercise. Certainly no two teams we observed on one day restored their grids in the same way. We take these different scenarios to mean there is equifinality in restoring the system. If so, then the fact that each scenario is very tightly coupled—literally line by line—contrasts sharply with the earlier argument that tight coupling offers few occasions for such equifinality (Perrow, 1999 [1984]). Second, authority followed function in the simulated recovery, something that has been observed in other high reliability operations (see Appendix 2). Even though formal authority relations exist between CAISO and the separate utilities, in practice an exercise of CAISO authority might have something to contribute at one point but not another, and when not, authority lapsed as a factor in team decision making at the distribution utility level.

To summarize, recovery was entirely dependent on the professionals working together, but their operations, both organizationally and cognitively, were very different from the performance modes we have observed when the system is running.

Another way in which recovery differs cognitively from resilience is that strategies undertaken during recovery may not feed back directly into the usable knowledge base of professionals. The hope of course is otherwise. Clearly, the training we observed during the simulation exercise was intended to increase learning and hone competence. Similarly, past recoveries have actually led to adjustments in reliability approaches through change at the macro-design level (for a brief history, see Apt, Lave, Talukdar, Morgan, and Ilic 2004). Most famously, the 1965 New York blackout led to a fundamental reorganization of the reliability system of the United States, including the creation of NERC, while the August 2003 Northeast blackout has led to making NERC voluntary standards now statutory.[5]

But at the same time, there are real limits to what a recovery process can actually add to the knowledge base of the reliability professionals, especially under actual operating conditions (and political pressures) and given the unique circumstances of each recovery (which is, as we have noted, an individual case). As one participant in the CAISO exercise noted, "We proved we could do it—in a simulation. But we don't know what an actual case would look like; for example what if we lost communications?"

NOW LET US TURN TO ROBUSTNESS. We frequently come across statements such as the following: "To work well, markets need both a robust legal framework and a behavioral infrastructure of accepted rules" (Plender 2002, 13). But just what does robustness mean organizationally, let alone cognitively?

As we reported in Chapter 5, one of the wraparound engineers told us about making systems more robust for control room operators. What was he talking about specifically? We are now in a position to distinguish reliability from robustness. As it turns out, the pursuit of robustness can lead to organizational conditions that undermine the cognitive processes of reliability professionals, especially their need to be resilient, and is therefore often at cross purposes with overall reliability. Indeed, in the way the term *robustness* is currently used, its application as a strategy to the systems we study could in fact seriously compromise their overall reliability.

Macro-design, micro-operations, pattern recognition, and scenario formulation have each been described as having its own kind of robustness. A common thread in each is that "robustness" ensures a system that does not fail. System designers may seek principles or rules that hold across different macro-designs at the system level (see Issing and Gaspar (with Tristani and Vestin) 2005, 6). They search out strong commonalities in macro-approaches that otherwise diverge. The most familiar example in the California electricity sector is the macro-design principle to promote robustness by ensuring that, whatever the policy mechanism for electricity provision—regulated, deregulated, or reregulated—it must enable increases in generation to meet load increases in the state.

One also finds robustness in scenario formulation, though of a different sort. We have observed the effort to move from macro-design to contingent scenarios by asking the question, Will these design principles work here, in this case or event, and if so, how is that scenario or protocol going to work specifically? This is what one policy analyst has called implementation robustness: "A policy alternative . . . should be robust enough so that even if the implementation process does not go very smoothly, the policy outcomes will still prove to be satisfactory" (Bardach 2005, 33). For example, a design principle (say, marginal cost pricing) might be robust across different regulatory regimes for the California grid. But it is quite another thing to insist that robustness requires a clear specification of scenarios for how that principle is to be implemented by the different distribution utilities—PG&E, SCE, and SDG&E—responsible for different segments of the California load on that grid and what potential impacts the implementation will have on actual consumer behavior there.

In our framework, reactive micro-operations can become the source of operator error. When an operator works outside the domain of reliability competence bounded by pattern recognition and scenario formulation, it is not surprising to us that uncertainty and the chances of operator error increase. Nor are we surprised that all manner of calls are made for "more robust technologies" to reduce the incidence of human error at the micro-operations level. It has become commonplace to recommend the introduction of advanced intelligent agent software throughout the California grid in just such terms: "In addition to relieving the operators from routine information management functions, the [intelligent software] agents will have advanced cyber-security capabilities and will employ secure multiparty cryptography to ensure robust operations immune to external cyber attacks and certain attacks from malicious insiders" (Smathers and Akhil 2001, 19). Here robustness describes how a given technology creates micro-operations that are supposed to be fail-proof; that is, without error or surprise.

Last but not least, we find another kind of robustness at the pattern recognition level. Is the pattern of observed behavior robust? That is, is the behavior systemwide and persisting? A common call today—and not just in the electricity sector—is to ensure that best practices are evidence-based. So too are there calls to ensure robust routines and standard operating procedures, a robust number of bids and trades, and robust demand growth in the electricity sector. In such usage, robustness is focused on the aggregated outcome across all cases, be they individual market transactions or energy trades. Thus no matter how error free or error prone behavior may be for individual transactions or trades at the micro-level, systemwide patterns across all transactions and trades can exhibit robustness when no aggregate failures arise, say, because of interaction effects. As an electricity expert told us in our 2001 round of interviews, "Traditionally we have looked at every failure we knew was possible. The problem is that there are hidden failures in the system—a protection device works one way when tested, but does not manifest certain modes until it is actually used in practice in interaction with other events."

In short, we have observed at least four uses for the term *robustness,* each addressing a different perspective on reliability. It is for this reason that we find robustness and reliability to be very different, for each type of robustness just described poses risks to the others. We believe that all these risks eventually will have to be managed by reliability professionals, if reliable operations are to be achieved.

No matter how robust the design principle, design principles have to be applied to case-specific contingencies that were never planned for, if reliability is the aim. No matter how robustly foolproof the micro-operations, these operations have to be examined for systemwide patterns, including unanticipated effects, if reliability is the aim. No matter how robust the case-specific contingency scenario or the systemwide pattern of behavior, they do not operate on their own, but must be sensibly connected in real time, if again reliability is the aim. Design protocols and established practices cannot on their own ensure reliability, without reliability professionals adjusting between the two when needed.

In brief, even robustness has to be managed reliably, especially so because of its multiple facets. We believe nothing is more challenging to reliability than the belief that robustness and reliability are one and the same thing. To believe that is ultimately to believe in the illusion of error-free design and surprise-free technology.[6]

WE CANNOT STOP PEOPLE from describing events in our case study in conflicting terms such as, "No, that's not resilience, that's robustness." Nor do we want to argue that such responses might not be merited in other studies. It may well be that, by the time all this terminology is sorted out, the literature will have moved well on from *anticipation, resilience, recovery,* and *robustness* to new terms, such as *agile, adaptive,* and *speedy* systems.[7]

That said, it is quite clear to us that the current terminology regarding anticipation, resilience, and robustness is misleading and could, if not thought through, lead to counterproductive interventions when it comes to reliability in the critical infrastructure we have studied.

CHAPTER 9

PUSH-PULL, REAL TIME, AND COST[1]

L ET THERE BE NO MISUNDERSTANDING: IF CAISO personnel had anything to say about it, they would never choose to be so exposed to highly volatile, real-time conditions on the order they have experienced since CAISO's inception. Control room dispatchers operate in just-in-time and just-for-now performance modes when they are pushed into doing so. "I learned about Murphy's Law," reported a senior grid operations engineer at PG&E. "If the worst can happen, it will happen. Which makes sense, because the worst is most likely to happen when the system is at its most stressed. Reserves are needed just when maximum transfers are happening." A National Research Council report (2002, 6-4) explains, "A highly stressed system (e.g., if power imports are high and transmission reserve capacity is low . . .) would be more vulnerable to cascading failures and the resulting longer-term blackout." Certainly, we know of no operator who is attracted to operating just-for-now. Chapter 5 described how operators are invariably pushed into that most unstable performance mode because of conditions beyond their control.

At the height of the California electricity crisis, a senior CAISO control room official compared his present situation with his past before restructuring:

To operate the system now, it's so dynamic. It's a full-time challenge. It's very volatile. . . . The biggest change are all the unknowns in all your decision factors and forecasts. A lot of decisions have to be made fast, in a short time. I have no guarantees that my forward schedules are right. I have no idea how much imports I can use for balance. There is a lot of intuition going on, and a lot of experience coming in.

It was more stable [at PG&E before restructuring]. At PG&E you had the generators under control. You knew the day before about today's capacity. Forecasting tools [were] better because you were dealing with a smaller grid and thus more precise. You knew supply and long-term deals that I don't have now. Some of us thought it was important to maintain long-term arrangements but the marketers wouldn't have any of that. We used to have arrangements with BPA. We would give them surplus in the fall and winter and get 1,000MW during peak during the summer. Now we can't do it. It makes our job a lot tougher. It's the biggest challenge in my career.

When we interviewed him again in 2004 he told us that the conditions he was then working under—new system peaks and just-for-now conditions—were even scarier than those of the 2001 electricity crisis.

YET FOR ALL THESE CHALLENGES, operators are also *pulled* to just-in-time performance, not just pushed there. In these cases, operating in real time confers advantages along with disadvantages (see also de Bruijne, van Eeten, Roe, and Schulman 2006). We have described how the tight coupling and complexity that are a threat in just-for-now performance can be a resource under just-in-time conditions, when equifinal options to balancing load and generation open up. What we observed were strategies to balance load and generation that were available in real time but not when the task environment was less volatile. In fact, these strategies worked better than some more conventional ones did when system volatility was lower. When exposed to high system volatility, such as rising temperatures, strategies whose very success depends on there being real-time conditions in place serve as pull factors to just-in-time performance. Seven of the factors are listed in Figure 9.1.

Before we discuss each factor, it is important to underscore the persistence of just-in-time behavior in the control room. We have observed and still observe the quickened pace of activities and many other features of "just-in-time" performance in multiple, repeated versions: in the early morning and evening when electricity usage increases considerably; in the late evening when generation goes off peak; during the ramps in between every hour; during the weekdays as distinct from the weekends; during winter months in comparison to the rest of the year;[2] during software upgrades and glitches in the control room that posed potential interruptions in service or grid reliability; and during scheduled outages that took place throughout the day, throughout the week, or over months. Real-time reliability is a concern every day, and so too are the factors pulling, not just pushing, operators to it.

1. Real time is an answer to persistent network incompleteness.

Part of the complexity of large technical systems is their set of multiple, at times conflicting, performance standards and measures. Because complex interorganizational networks can never be fully designed beforehand, considerable weight ends up being put on real-time activities to ensure the reliable provision of services. In real time, almost everything that matters for reliability is denominated in terms of load and resources, while performance standards are

---

1. Real time is an answer to persistent network incompleteness.

2. Positive interdependencies are most evident and likely to be acted on in real time.

3. Real time remains informal, nonroutine, and flexible.

4. Real time allows for larger process variance relative to output variance around the balance of load and generation.

5. Real time legitimates and accommodates the redefining of reliability criteria.

6. Service reliability is fixed and given, except when risking grid reliability in real time.

7. Real-time operations justify and reward improvisation and experimentation.

FIGURE 9.1. Selected pull factors to real-time reliability

their clearest when it comes to staying within the ACE bandwidths or violating them. There is incredible pressure to pull the electricity network together at the last minute when success or failure is most visible and salient to everyone concerned. In the words of one of CAISO's lead officials in marketing operations, "in real time everything comes together here in the control room." Also, it is not surprising that the operator leaves some important decisions to the last moment, when information has the greatest probability of being timely and reliable. Fortunately, although system design remains incomplete, efforts have been made, most notably through Resource Adequacy and other changes, to increase the stability of network relations.

2. *Positive interdependencies are most evident and likely to be acted on in real time.*

None of the other control areas in the West can afford an unreliable California grid. Most immediately, instability in the state's grid (including plant and transmission lines) can ramify and cascade into instability throughout the wider western grid.

In this setting, positive interdependencies come to the fore when needed most. A senior CAISO control room official pointed out that in one month during the electricity crisis Bonneville Power Administration (BPA) bailed CAISO out numerous times at the last minute, because BPA knew that California would be needed to go the extra mile for them later. In his words, "the rest [of the western grid] can't afford California to go down. BPA knows that it needs us in the fall." There are times when different control areas are chasing the same megawatt, but positive complementarities between control areas are their most palpable and evident in real time.

There are no guarantees here, however. A year after our interview with the senior control room official, he and others informed us that it was increasingly difficult to depend on adjacent control areas to bail them out at the last minute with imports. As we saw subsequently, the dominant performance mode increasingly became just-for-now, in which low options variety made it very difficult for CAISO to reciprocate with power to out-of-state grids. That said, the virtues of interdependencies have not disappeared and may have been strengthened, as we saw in July 2006, when CAISO broke its load record.

3. Real time remains informal, nonroutine, and flexible.

When incompleteness in formal design reinforces the importance of getting operations right in real time, many of the control room operational activities in real time have understandably remained informal and tacit.

Real-time reliability in the control room places great value on the nonroutine over the routine, the informal over the formal, and the relational over the representational. Much is handled hour by hour and best summarized in the phrase, "operators using their discretion." A senior manager in CAISO operations engineering unit, responsible for a large body of procedures, told us during the electricity crisis, "part of the experience is to know when not to follow procedures. . . . there are bad days when a procedure doesn't cover it, and then you have to use your wits." Smart people use their discretion, and it takes very smart people—in fact, professionals in high reliability management—to run the control room in order to produce grid and service reliability in real time. Design-induced glitches happen, and the role of the reliability professionals is to work around and manage those glitches.

The mix of informal, tacit, nonroutine, and relational is best seen in the shifts to and from just-in-time performance under high system volatility. We saw in earlier chapters how the control room tempo and pace quicken under increased volatility, sometimes to the point at which emergencies are declared and the control room team widens considerably to include an emergency response team of experienced engineers and higher-level officials who have working relationships with large importers of electricity into California. The capacity of key players to make informal mutual adjustments within the control room, its surrounding support staff, and the restructured network elsewhere is a core part, we found, of ensuring real-time reliability.

We cannot overstress the importance of teams and flexible authority patterns in ensuring reliability and how appealing this is to reliability professionals. A team involving different positions and roles is in effect a matrix whose configuration can go lateral or vertical, depending on what is required in real time: one minute the shift supervisor is treated as colleague whose advice matters to the operators, the other minute he or she is called up to make a decision or to buffer the operators from outside interference. The teams are able to take on a matrix structure precisely because hierarchy gives way to experience and skill when real-time reliability is at stake. Authority, as we saw in the last chapter, follows function. But we should never lose sight of the fact that the matrix works because going lateral carries the possibility of going vertical in order to ensure grid and service reliability. "Sometimes I talk to the generation owners," a CAISO shift manager observed early on in the research, "but only if they don't do what we need them to do [in order to ensure grid reliability]. Bully them if needed. . . . Sometimes I pull a little rank, a little authority."

4. Real time allows for larger process variance relative to output variance around the balance of load and generation.

The operational requirements of the CAISO control room are so intense as to be single-minded. Operators and their immediate support staff are focused on overarching reliability concerns to such a degree that everything else is "politics."

But although the output of CAISO is clear—reliable electricity—the options and resources to achieve that objective are often diverse. This has come about because there are many more interventions, transactions, and contingencies in the control room after restructuring than before in the older utilities. A sense of this shift to wide process variance around a committed low output variance is captured in the 2001 comments of a senior CAISO scheduling manager:

The old way was nicer and easier, you had twenty years of experience behind you, but now you still get the [same] results. The fact that a lot of it has gone into real time has caused headaches, but that's it. There was more tunnel vision before, scheduling was a narrower task. It has widened out, it encompasses more, pushing things into real time, there was never before a BEEP person, there wasn't an hour ahead, the tunnel is widening a bit more, but concepts remain the same. . . .

This large variance can work against achieving reliability—many more things can go wrong now—and there will be strong pressure for CAISO, from within and without, to reduce the number of variables with which it has to deal. Yet the large variance and single focus of CAISO also pose opportunities for real-time reliability in several ways.

First and most obviously, the large variance represents equifinality in achieving reliability under just-in-time conditions. As we have already noted, operators see the connection between having more variables to worry about and having more options to come up with possible solutions, whether in the market or out of market.

Second, the single-minded focus on reliability means not only that the performance standards are their clearest in real time but that feedback on whether these standards are being met is almost immediate in terms of frequency, ACE, and CPS2 violations. That is, feedback is clearest about whether the service is "on" or "off." The reliance on real-time feedback and information, in turn, reinforces letting performance remain "just-in-time."

How? Operators and engineers in the control room can look for and read those signature events around load and resources discussed earlier, while admitting they do not fully understand or comprehend what is happening in the entire system at that moment. In fact, just-in-time performance would not be "just-in-time" if there were not some lack of clarity about what is going on in the network and wider electricity system. In the words of a WSCC (now WECC) security coordinator stationed in the CAISO control room, "Ninety-nine percent of the time things don't happen in real time as scheduled." "Generators are rarely where they are supposed to be. That's the nature of the beast," said a CAISO engineering manager during the energy crisis.

Third, large variance in terms of control operators' experience is a resource when the unexpected and contingent are "normal." After describing the wide background of his control room crew, a CAISO shift manager in 2001 summed up by saying that there was two hundred years worth of experience there in the room. Another control room operator in CAISO told us that the wide experience meant that it was much more likely that someone in the control room knew how to handle or deal with something surprising. Wide experience, combined with formal and institutional knowledge of control operations by operators (often achieved through training on the job and over multiple positions), has served to promote and reinforce reliability, particularly when performance urgency is high.

Large process variance also comes in the form of experience of the wraparound support staff. Some operators in the control room said they scarcely noticed the demise of the Power Exchange, the state's primary energy scheduler, the day after it went bankrupt. A variety of people in the wraparound (including those in market and grid operations) scrambled to ensure that new rules and procedures were in place to make the transition as seamless as possible. Consequently, much of the wraparound takes on real-time concerns in its core activities, be it outage coordination and prescheduling on the one end or settlements on the other.

But as we saw in Chapter 5, the experience pool, both in the wraparound and in the control room, has changed and diminished since the electricity crisis. By how much we cannot determine. Recent efforts at increased cross-training of positions as well as other exercises (for example, the black-start recovery simulation) may have compensated for some loss. Similarly, the control room remains a crucible for building up real-time patterns and scenarios for new recruits, as periods of high system volatility continue up to the present.

5. Real time legitimates and accommodates the redefining of reliability criteria.
One of the advantages of real time is that certain existing standards, such as those governing the air quality of generators, can be relaxed and thus can open up more options for operators to ensure real-time reliability. Indeed, as a result of restructuring and the electricity crisis it induced, *new* reliability criteria and standards emerged. The pressure has been to adapt reliability criteria to meet circumstances that operators can actually manage, when those circumstances have been increasingly real time in their urgency. Again, what cannot be managed through one performance mode is managed through another, or, barring that, through just-this-way by shedding load directly. Having standards that can be met, albeit in real time, has great appeal to operators.

Consequently, there is the paradox between having firm reliability standards and having multiple ways to produce electricity reliably that we mentioned before. On the one hand, the standards are operationalized by NERC and WECC, and because performance can be empirically gauged against these operational measures, they become the chief measuring stick of whether electricity is being provided reliably or not. On the other hand, and to reiterate the point made in Chapter 4, the standards were everywhere under pressure in 2001 and later. Only by pushing the standards to their limits and sometimes beyond were the lights kept on in a number of cases.

Thus, just as we can expect the network to remain incompletely designed for the foreseeable future, so too can we expect the standards to be continually under pressure of real-time modification by the professionals who operate our critical infrastructures.[3] Certainly the federal legislation that followed in the aftermath of the August 2003 blackout in the Northeast brought an entirely new set of complexities to the table, formalizing into law what were once voluntary standards and processes of standards making. We can expect the pressures to modify these standards in light of real-time conditions to persist.

6. Service reliability is fixed and given, except when risking grid reliability in real time.

To put it succinctly, control room operators and engineers are pulled, not pushed, to real time by the fact that in real time there are no substitutes for high reliability in the balance between load and generation. If load and generation are not equated in the last instant, the issue of servicing load is moot. There will be no electricity unless that balance is struck and, accordingly, control room operators treat grid reliability as nonfungible when it matters the most, right now.

As one senior state government official put it to us in 2001, electricity "is nonsubstitutable. Yes, it's possible to use gas for some machines, but that is limited. It's not like going into a bakery and wanting French bread, but choosing bagels. In the case of electricity, the only substitutability is the decision to consume or not to consume. . . . " The properties demanded of electricity—its large-scale provision must be safe, continuous, and available—are treated as an unavoidable high reliability mandate. In the 2001 words of a state legislator specializing in energy policy, "When we look at how we use [electricity] we see there isn't a very good substitute. It is essential for water purification, hospitals, a whole host of communications, it can't be stored, it has to be delivered in real time. . . . "

In this way, electricity is an essential given that cannot be traded off, at least in the very short run. Indeed, service reliability is given up only when it directly jeopardizes grid reliability. This is an attractive position for operators and engineers to be in, particularly when the high system volatility they are operating under is none of their creation but balancing load and generation is their entire job. Given long enough, of course, all fixed costs become variable. In the long term, fixed inputs give way to substitutes. But real time is not the long run. Indeed, a good definition of real time is that point in critical service provision when there is no substitute for the high reliability management of our critical infrastructures.

7. Real-time operations justify and reward improvisation and experimentation.
As we have detailed, nuclear power plants, aircraft carriers, and air traffic control systems have avoided for the most part anything like the large-scale improvisation we have found in the California case study. What was unacceptable in the older vertically integrated utilities has become *sine qua non* for service and grid reliability in the restructured electricity network—but only in real time.

We have already discussed why operators have had very little choice in making this experimentation systemwide in real time—at best they undertake involuntary design probes of the grid. We also saw in the preceding chapter how recovery pushes operators to a scenario formulation that operators would rather have foregone. Improvisation-as-operations has, however, occurred for pull reasons as well. Technological improvement in electricity provision continues to be a major driver of engineering-induced change. The electricity sector has recently accelerated its search for better software, equipment, and approaches to real-time grid management. A senior grid operations engineer at PG&E told us, "We now have a menu of analytic tools that we never had. You find out things that you never thought you would find out otherwise. We now operate closer to the margin, more economically, with better utilization of equipment or installation of equipment."

Engineers are trained to innovate when confronted by persisting problems (von Meier 1999), and engineers dominate CAISO's wraparound. "Engineers always like to do new things," a CAISO transmission planner told us. Said a CAISO engineering manager about his support work for the control room, "It's pretty pioneering." Said yet another grid operations engineer, this time at PG&E, "I like pushing systems to the margins, but not one inch beyond. That's why we need to have that real-time capability, that knowledge of the system; having all the data coming in and the ability to crunch the data."

OTHER FACTORS pull operators to real time, but our point is made by the seven just discussed: real time is a persisting feature of CAISO's high reliability management within the control room, not just for one reason, but for many. Not just because it is all push, but because there are also pulls to real time. Nor are these factors important in CAISO alone. De Bruijne (2006) has found many of them to be critical as well in explaining the real-time reliability of the major Dutch mobile telephone system, an industry that also has been subject to competing and fragmented interests.

One of the major consequences of these pull and push factors to real time has been to keep cost issues a live concern in the control room and within CAISO generally. In the popular mind, the impression has been that high reliability means reliability whatever the costs; indeed, the nonfungibility of grid reliability in real time would seem to support that conclusion. In reality, the pull factors of real-time nonfungibility, redefinition of reliability standards, and system interdependencies have combined with other factors to keep cost and cost reductions an ever-present and pressing issue at CAISO. Because cost is an important issue in itself, it is worth discussing this topic in detail.

Start with the issue of nonfungibility and the cost of electricity in real time. "This is system reliability," as a shift manager put it to us in 2001. "They tell me on the phone that it's expensive. That's not really my concern. I don't spend unnecessarily. But reliability goes first with us."

"I don't spend unnecessarily." What does that mean? Such considerations of financial costs have been repeatedly mentioned in our interviews, inside and outside the CAISO control room, and the considerations invariably relate to cost-effectiveness: Is high reliability being provided at low or minimum cost?

Over the years, we have heard frequent complaints about generators gaming the system, with the added settlement costs passed on to consumers. We have also often heard within CAISO about the need to keep its management costs low, if only to justify CAISO transmission management to the utilities and energy suppliers. Moreover, we have found many instances of efforts to reduce the costs of operation. The cost savings that the CAISO control room tried to realize, for example, include looking for uninstructed deviations in the Automatic Generation Control (AGC), as they can mean larger settlement fees.

Such cost savings, however, have been sought as long as they did not compromise reliability. This notion of high reliability at its least cost is not new. Under the older cost-of-service reliability of the integrated utilities, "we chose the standard of reliability for our customers and then estimated how to do that in least-cost terms," according to a utility director of a maintenance and construction division we interviewed in 2001. The "we" were engineers in consultation with operators who developed the standards, while the estimation of cost was left to the economists.[4]

But although reducing the costs of grid management has been an explicit concern of many CAISO staff interviewed over the years, electricity restructuring has transformed the older calculus of cost-effectiveness. It is critical to understand this transformation and its real-time implications.

First, the financial costs of electricity transmission became more important than they ever were before as a result of the electricity crisis, and they have always been important. CAISO officials walked away from the crisis in agreement: never again. If they had any say, never again would CAISO find itself keeping the lights on without regard to cost. To have repeatedly warned the regulators about the market manipulation and then, when nothing was done, to find themselves the conduit of the nation's largest public-to-private, state-to-state (from California to Texas) income redistribution scheme ever left the taste of ashes in the mouths of CAISO grid managers. One cannot understand CAISO's subsequent efforts ranging from proxy bid marketing, through must-offer requirements, to Resource Adequacy initiatives without seeing them as deliberate attempts to bring cost-effectiveness and its realism back into real-time reliability management.

Bringing realism back through the calculus of cost-effectiveness has been made all the more important because of the trade-off between cost and reliability. As we just discussed, what is fixed in the immediate term becomes variable over the longer term. Our critical infrastructures and their levels of reliability, which managers are to manage in real time, are developed and utilized over time in light of opportunity costs. But that trade-off between cost and reliability existed more in theory than in practice during the electricity crisis. In 2001, no one in the restructured electricity sector had the duty or responsibility to decide that, "If the price of the next megawatt is $X, we will declare blackouts rather than purchase that megawatt in order to keep the lights on." Not CAISO, not the CPUC, not the distribution utilities, in the end not even the state itself turned out to be in a position to take responsibility for turning the lights off to stop the financial crisis into which all rushed like moths to an open flame. What has been subsequently happening through Resource Adequacy and related efforts is an attempt to ensure that someone in the state take that decision when needed. Such efforts, in other words, have been to make the trade-off much more real—and real time—than it was in 2001. A 2007 conversation between a gen dispatcher and shift supervisor catches the trade-off nicely. The shift supervisor was explaining some of the advantages promised by the forthcoming MRTU (Market Redesign and Technology Upgrade). "Right now, say we don't care between choosing [two specific generators]. What MRTU will do, and it's a great thing, is to choose the cheapest between them." To which the gen dispatcher quickly added, "As long as it is the one who historically comes to the game!"

The older calculus of cost-effectiveness has been transformed also because of the fact that standards for reliability were tested and changed as a result of the crisis. Once standards are called into question or new ones emerge, the determination of cost-effectiveness necessarily becomes more complicated and ironically even more an issue: Just what standard of reliability are you going to hold constant so as to minimize its cost? In these circumstances, it came as no surprise that differences between the regulators and CAISO were to be expected and highlighted, depending on the standard used in the cost-effectiveness calculation. A WSCC (now WECC) security coordinator described one such case to us during the crisis: "It's a NERC regulation that every generator that is online is supposed to be under AGC [Automatic Generation Control]. CAISO doesn't operate like that. . . . Why? Because it is not cost-effective."

The calculus of cost-effectiveness also changed in another way. Cost-effectiveness cannot just be defined as least cost for a given level of reliability. There is also the other, better known side of cost-effectiveness: the biggest bang for the buck. In other words, for a given cost, what is the highest reliability that you can manage? In this cost-effectiveness calculation, the roles of economist and engineer are reversed. The economist determines the cost, and the engineer is expected to provide guarantees of the level of reliability that can be optimized for that cost. This reversal, of course, is a much more serious challenge to high reliability management, as there is no up-front guarantee that the baseline cost will ensure baseline reliability. Cost considerations thus again are forced to move center stage, particularly as the real-time fear now is that considerations of marginal reliability will altogether replace those of precluded events reliability, so important for high reliability as we saw in Chapter 3.

We certainly have seen pressures to accommodate considerations of marginal reliability in ways that are said not to compromise high reliability management. We see this in its 2006 business plan and associated statements that CAISO will no longer be the buyer of last resort, albeit high reliability is to remain the priority. We also saw it in the electricity crisis, again in the shifting standards issue. As a senior CAISO engineer told us in 2001,

All performance standards are being opened up. The fact that you have to pay for it is why they are being looked at. The fact that the standard "felt good" is not good enough anymore. And some will say, "dammit, it always worked and you are eroding reliability now." And they are probably right, but eroding it by how much? Probably not by much, and that little bit of reliability is bought at a high price. So it is an acceptable risk, and it is even desirable to change it.

What mitigates this threat of moving to marginal reliability is another part of the calculus of cost-effectiveness that is often missed: high reliability management can actually be the basis for *reducing* costs; conversely, cost reductions can actually *enhance* high reliability management. Either way, it would be a mistake to conclude that cost reductions for any reason are the enemy of high reliability. Or to put it differently: a positive relationship between cost and reliability forces us to rethink that longer-term reliability-cost trade-off.

Clearly, the drive to reduce costs and ensure, if not enhance, reliability is seen in CAISO's adoption of dual-use technologies and management strategies that work to improve both normal operations and emergency response. These are ways to be reliable and cost-effective at the same time and for real time. They include cross-training of a multiskilled staff, off-site backups and computer servers, simulators for training operators as well as pretesting new software, flexible contracts with suppliers, and technology and software, such as a state estimator, that searches for patterns as well as generates contingency scenarios in the absence of patterns (for more on dual-use technologies, see Sheffi 2005; Allenby and Fink 2005). In this way, the technology and management that the critical infrastructure would have had to pay for anyway enable it to provide other services without degrading real-time reliability and even enhancing it.

Second, examples abound in which, after given levels of high reliability management are achieved, a platform then exists for reducing costs. Indeed, we believe this to be the primary way an organization such as CAISO becomes cost-effective in terms of finding the least cost for a given baseline of reliability. Moreover, the baseline is raised when the cost savings are actually reinvested to improve reliability management.

Here we come back to that other pull factor, system interdependencies. The networked reliability discussed in this book is largely a supply chain of generators, transmission, and distribution utilities, in which the overall reliability of the supply chain is also a precondition to each of its elements being productive. The goals and objectives of each element can well be conflicting, but without the baseline reliability of moving the electrons along the supply chain from generator to end-user, none of the goals and objectives would be realized. Cost reductions of any element in the supply chain may be invested back into enhanced baseline supply-chain reliability as a way to make that element even more productive. Certainly that is the hope with respect to Resource Adequacy requirements, in which the rules of the game are much clearer about who is to supply what, when, and where within and along the supply chain. It is anticipated that, once these rules are in place, more effective and efficient transmission planning as well as coordination of generation construction and outages become possible.

However, these anticipations are just that, and must always be balanced by real-time resilience strategies of elements along the supply chain to respond to the unforeseeable. Indeed, this balance of strategies is required not just *in spite of* the restructured design but also *precisely because of* that design and continuing redesign efforts that have followed. We take up the importance of design in the next chapter.

DESIGN, TECHNOLOGY, AND RELIABILITY CHALLENGES

A S THE PRECEDING CHAPTERS HAVE SHOWN, significant differences exist between what is commonly believed about risk, resilience, anticipation, robustness, and real-time operations in critical infrastructures and what we found to be the case for the California electricity grid. So too do we find differences between expectations and reality when it comes to the role of technical design in infrastructure reliability. We believe many stakeholders, including regulators and legislators, confuse the macro-design orientation with the actual impact of formal designs, whether in law or policy or new technology, on our critical infrastructures. Formal designs and new technologies can improve reliability, *but only when they are managed to achieve that end.*

In the world of high reliability management, design errors in policy or technology are inevitable despite the most careful efforts. In California electricity restructuring, in spite of the stated and restated conviction of policy

architects—economists, regulators, and legislators—that electricity markets
would quickly evolve and attract new players in electricity generation as well as
keep wholesale prices down by market forces, this utterly failed to happen.
Managers of the grid confronted quite different conditions and had to keep
the lights on nonetheless.

IT IS EASY FOR OUTSIDERS to think that macro-designs govern reliable performance, especially when they have no idea about the role and domain of competence of reliability professionals who translate design into practice. It is vital, however, to differentiate the macro-design orientation from the
orientation of reliability professionals in order to understand how design and
actual operations diverge.

The macro-design orientation does not disappear in the realm of reliability professionals. After all it provides the principles that must be applied and
modified case by case through contingency scenarios. That said, those processes
of application and modification should not be confused with macro-design
and the orientation of designers themselves.

The two orientations embrace different values. In high reliability management, ambiguity can be strength. Much is accomplished by good judgment
rather than explicit instruction. Ambiguity allows discretion and protects against
error in explicit commands. Whereas macro-designers value clarity, precision,
and consistency (Pool 1997), managers value judgment, discretion, and trust.
In the world of high reliability, judging each case on its own merits can be the
very best management possible. Yet all too often the complaint by outsiders is,
as one CAISO stakeholder put it, control room "dispatchers schedule the system using experience and mental models that are not precisely defined and
available a priori" (Sibbet and Lind 2000, 18).

Each orientation also requires different skills and outlooks. The world of design is a province dominated by lawyers, economists, engineers, model builders,
and technical consultants—what organization theorist Henry Mintzberg (1979)
has called the "technostructure" of an organization. The managerial and operational levels, on the other hand, require a different set of cognitive skills.
Managers frequently have to balance competing principles or logics (such as
the trade-offs when protecting against errors of omission versus errors of commission) and buffer contradiction or paradoxes associated with multiple design values (Rochlin 1993).

The approach to error differs as well between the two groups of professionals. Designers often underestimate the prospects of human error or, at the other extreme, are committed to the elimination of error through anticipation and proscription. To that end, engineers value innovation, as we saw earlier. But managers value skepticism and simplicity. "It is never proof in the pudding until you see it work, until it runs," said a gen dispatcher. "Operators don't want all this noise; 'just tell me the stuff I need to know' they say," a wraparound engineer told us. "It's a common theme of their jobs—the number of procedures have increased, the procedures have gotten to be more complex, how can we make this simpler, there are so many things to remember, and how can we simplify?" (see also von Meier 1999).

Operators approach the design and redesign of complex systems cautiously, fully expecting mistakes and failure with any innovation. Good management is the management of error, especially the balance of error tolerance and error intolerance, as we have seen. In fact, a major function of reliability professionals is safeguarding operations from errors at the design level, whether the errors are from legislatures, regulatory agencies, or the engineering departments of their own organizations. In this way, reliability professionals are a vastly underappreciated societal resource.

Designers and high reliability managers are also oriented quite differently toward failure probabilities and their analysis. When reliability professionals express their "gut feeling" or concern for the reliability or safety of a system under management, engineers and designers often insist that these concerns be expressed in terms of a formal failure analysis, which shifts the burden of proof to those concerned to show in a specific model the ways (and probabilities) for a failure to occur. This approach rejects the tacit knowledge base from which a pattern might be discerned or a failure scenario imagined.

Ironically, the very same engineers may at the same time assert that the reliability of the system they have designed is based on past operational experience. If a technology has worked in the past, so too is it likely to work in the future. This asymmetrical treatment of knowledge—requiring prospects of failure to be formally modeled, but accepting a tacit, experiential *post-hoc* confirmation of reliability—can be seen in NASA's treatment of safety concerns about the Columbia shuttle after it was struck by tiles during launch (Starbuck and Farjoun 2005; Vaughan 2005). The retrospective approach of the design perspective—that reliability can be established on the basis of a past record of successful performances or deduced on the basis of prior principles—contrasts sharply with the high reliability management discussed here.

For example, an engineer in the CAISO wraparound took issue with the criticisms of a control room official over a state energy assessment for the period ahead: "I don't know what he's upset about. Last year we got through. Yes it would have been difficult if it had been a one in ten year, but it was a one in two year and we did get through. Yes, his list [of concerns] includes generators without firm contracts, but this is the same list that provided us generation last year without any trouble. There's no reason to think that what worked last year won't work in the next one, assuming it's not a one in ten year."

In contrast, high reliability management for control room operators and staff is a prospective orientation. You are as reliable as the first failure out ahead of you, not the many successful operations behind. High reliability managers are always "running scared" with respect to the problems they are likely to encounter and the ones they have yet to foresee. But they do not rush to conclusions.[1] Their aim is to avoid letting mistakes happen or to work around the ones that have actually occurred in order to ensure reliability. To do what they need to do, managers have to combine analysis and judgment, a practical craft that resists retrospective validation, let alone prospective formalization. In an insightful analysis of NASA's earlier shuttle disaster, the Challenger explosion, Diane Vaughan, a sociologist, came to an important conclusion that bears directly on the contrasts we have been describing.

I had micro-level data on NASA decision making and also macro-level data on the NASA organization and its relations in its political/economic environment. These data made it possible for me to explain how and why signals of potential danger [regarding the shuttle] were misinterpreted. The answer was "the normalization of deviance": a social psychological product of institutional and organizational forces. The consequence was that after engineering analysis, technical anomalies that deviated from design performance expectations were not interpreted as warning signs but became acceptable, routine and taken-for-granted aspects of shuttle performance to managers and engineers (Vaughan 2005, 35).

A retrospective reliability perspective coupled with priority on formal analysis can actually propagate the deviance Vaughan describes. Yet sensitivity to these deviations is precisely what high reliability management is all about.

Deviations from a macro-design approach must be expected; indeed that is why the formulation of contingency scenarios is so important when thinking through how design principles are to be applied to the case at hand. From a manager's perspective, deviations are potential precursors to system failure and thus demand all the sensitivity and skills that the manager has in scenario formulation and pattern recognition.

With its prospective reliability focus, high reliability management is very attuned to precursor conditions that challenge or threaten future performance. High reliability organizations identify a precursor zone of operating conditions or circumstances they wish to stay out of, as discussed in Chapter 5. This zone is carefully defined and analyzed, *both* formally and experientially (Schulman 1993b; National Academy of Engineering 2003; Carroll 2003). The precursor zone in high reliability management highlights anomalies; it does not "normalize" them.

THE DEVELOPMENT OF OPTIMIZING MODELS for controlling infrastructures underscores the very real differences between the design and managerial orientations and why these differences pose special challenges to reliability professionals. Many optimizing models are designed to utilize capacity to the fullest, or to produce output at the cheapest cost or to capture the greatest market profit or share. The models may scan hundreds of variables—such as all available generators in an electrical grid; market prices of their power; transmission line conditions, including all relays and switches; all alternative line configurations; comparable costs of power imports; upcoming maintenance schedules; and the like—for an optimal performance "solution" under which to direct operations. These models may also continually recompute these solutions over short time intervals (in seconds or minutes) in order to pursue shifting optima or maxima. "The brain is inside the computer now," a senior CAISO market operations said of his unit's software in 2004.

A problem arises when the models provide solutions that are increasingly difficult if not impossible for control operators to follow or track in real time (Chapter 5). The speed and complexity of the calculations if coupled to automatic dispatch or control systems can place an infrastructure's operation effectively beyond the cognitive competence of human beings, including reliability professionals. Their ability to rescue the situation if errors or surprises attend the operation of these nontransparent models is degraded.

When we were writing our January 2005 memo for the CAISO COO, we were especially struck by this gap between cognitive skills and the requirements of new software and technology, particularly with respect to the replacement of the BEEP desk by the RTMA software. It seemed to us that the following assumptions would have to first be validated before one could call what we were observing in late 2004 high reliability management:

- New tools and technologies will provide more options without simultaneously increasing the volatility (that is, uncontrollability and unpredictability) of the system.

- Input errors into new software systems such as RTMA will be detectable by control operators and will not degrade their accurate pattern recognition and ability to develop anticipatory scenarios.

- Faster transaction speeds, promoted in new software such as RTMA, will not propagate error. Whatever the speed of the transaction, it can be assessed and comprehended by control operators and errors detected and corrected before downstream consequences build upon them.

- Operator skills that are not used will not be at risk for atrophy. Operators will not become complacent and assume that "the software must know what it's doing."

- Operators can perform for extended periods in just-for-now performance conditions (in which a small number of control options confront significant volatility in grid conditions) without failure in grid or service reliability.

- Increased complexity and tight coupling of the grid, combined with more unfamiliar conditions, do not degrade pattern recognition and contingency scenarios among the operators and their wraparound support staff.

- The skills or technologies available to operators will be able to compensate for errors induced by the pressure of just-for-now performance.

- Development of procedures is a cumulative increase in the knowledge base for grid management, and procedures are understood, recalled, and acted upon by control room operators and, when necessary, wraparound support.

- Finally, if automated systems fail and earlier operator skills have to be activated, this can be accommodated by reducing transaction speeds in order to match operator skills and reaction time.

We were not confident then nor are we confident now that these assumptions hold, and not just in the critical infrastructure of CAISO. The issue here is to ensure what could be called cognitive neutrality: new variables are introduced into high reliability management only if they pose no net increase in the cognitive requirements of their control room managers. Otherwise, new designs and technologies that are not cognitively neutral but rather add cognitive burdens for reliability professionals end up threatening the pattern recognition and scenario formulation so key to high reliability management.

JUST HOW DO ILL-CONCEIVED macro-designs and technologies attenuate the skills of professionals to recognize patterns and formulate scenarios? How does bypassing reliability professionals make things worse for high reliability management?

The answer takes us back to the problems associated with the just-for-now performance mode. As observed in the CAISO study, just-for-now performance becomes increasingly unstable the longer operators persist in firefighting and quick fixes, where doing one thing makes another worse off. Eventually operators come to be working in areas that are unstudied conditions for them, where few if any patterns are recognizable and few, if any, scenarios have been or are being formulated. If such performance is prolonged, operator errors and system failures become inevitable, ironically requiring what looks to be remedy in better macro-designs (Figure 10.1).

The dynamic works this way. Under persisting just-for-now conditions two major threats to high reliability management arise. First, the domain of reliability professionalism shrinks. As operators spend more and more time firefighting and undertaking quick fixes and patching up the system with what they know to be band-aids of all sorts, they cease to be professionals in the same creative ways they were in the other performance modes. Second, the longer they remain in just-for-now, the more they are expected to perform reliably in areas for which they cannot recognize systemwide patterns as well as for which they have few if any local contingency scenarios. In either case, operators are asked to use their best judgment precisely in those situations and under those conditions in which judgment is least reliable (see Hammond 2000). Mistakes are inevitable. Instead of starting from macro-design and micro-operations and tacking to the domain of professionalism, the pressure now

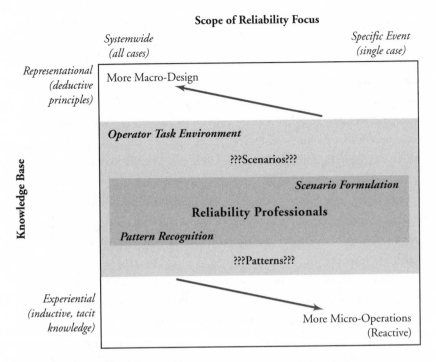

FIGURE 10.1. Prolonged effect of just-for-now performance mode

is to end up with more macro-design as the answer to more consequential micro-errors at the operator level.

In short, macro-designers must be very chary of cognitively burdensome assumptions about how much real-time reliability performance and associated wraparound behavior can be formalized or routinized, given volatility that cannot be controlled or predicted and given the adaptive, last-minute nature of the equifinality involved in high reliability management. The balance between adapting to the unpredictable yet having formal rules in case the worst becomes unavoidable is reflected in remarks of a PG&E shift supervisor we interviewed in 2001:

Out of the six [blackouts], I've done four. . . . It's been a changing, living document. What's been happening has never been done. So we're trying to write rules for something that's just happened, but each rolling blackout is new. It was a nightmare.

In this kind of world and from the perspective of high reliability management, any new design or technology must, if it is to take reliability seriously, pass the reliability test of reducing system volatility and/or increasing options variety and/or enhancing the cross-performance maneuverability of operators to respond to changing conditions in both.

Yet macro-level design seems perversely to move in the opposite direction. In fact, it has been our observation that macro-designers are tempted to see the domain of competence of the reliability professional as something on which they do not want to rely. They would rather not trust that expertise and discretion. They see bandwidth management under active analysis and think the answer is to design a fully anticipatory bandwidth management within prior analysis. They see multiple bandwidths, and they want to optimize them into zero bandwidth with no adjustments beyond those specified in principle, precedent, or theory.

In our view it is sheer hubris to attempt to shrink bandwidths by political fiat, legal rule, or regulatory punishments that do not rest upon a firm foundation in settled knowledge based on the mix and interaction of formal and informal understandings of the large technical system in question. To disregard the unique knowledge of reliability professionals would be to build control frameworks upon illusion—hardly a recipe for enhancing the performance reliability of any organization, let alone society's critical infrastructures. Certainly the restructuring of California's electricity sector suffered no management realism to darken its design.

Yet many system designs and redesigns are conducted within this same realm of illusion—a world in which unified networked designs ensure consumers ubiquitous service coverage, easy access, and throughout high reliability—and these have left our critical infrastructures less reliable and more vulnerable. We believe that these illusions are the real challenges society faces when it comes to the high reliability management of critical infrastructures. The illusions are, moreover, rooted in a broader policy context that increasingly sets what we see as a hostile environment for critical infrastructures and their management.

LOOK FIRST AT THE LARGER CANVAS of public policymaking. A growing number of public policy designs feature one or more of the following characteristics and assumptions.

   1. Problems or conditions to be addressed by policy are defined primarily in formal, abstract analytic models rather than widely understood pragmatically on the basis of experience.

In this respect, policy design can be far removed from the frame of reference of everyday experience, addressing problems founded more on generalization than on descriptions of existing conditions (Scott 1998). They may address problems such as ozone depletion or global climate change that are primarily identifiable through statistical correlations or the outputs of sophisticated modeling, rather than perceivable though direct experience. New policies may center on the creation of things that are artificial—new and complex technologies (such as a new weapons system), new markets (such as emissions permits or the electricity markets we have been describing), or even new forms of organization (such as mixed public-private "governance" frameworks). Today, a great deal of public policy design is associated with novel approaches to institutional frameworks for governance, such as public-private partnerships, coordinated contractual relationships among multiple organizations, or collaborative networks of diverse interests and stakeholders (Imperial 2005; Salamon 2002; Savas 1997).

2. Key design variables and policy performance standards are formulated thoroughly in artificial language rather than in the natural language of everyday use.

Key concepts, such as "optimum" or "market clearing price" are defined in mathematical or game theoretic terms. Regulatory standards are also formally defined, such as in "parts per million" of a particulate in a given volume of air, as an air pollution standard, or in the CPS1 and CPS2 standards applied to power grid management. A lowered tolerance for ambiguity attends the formulation of modern policy. It is a hallmark of the contemporary to spell it out, nail it down, and, above all, get it in writing.

3. Policy design is directed toward the extreme case (either the worst-case scenario or an optimized or maximized case) rather than an average case or run of cases.

The archetype for this design approach is found in nuclear deterrence policy (Kaplan 1991). Here the idea has been to assume the worst in both enemy capacities and intent, and to design an offensive arsenal "robust" enough to withstand first-strike attack. More recently, policy designs have been directed toward seeking maximal or optimal output, that is, the most cost-efficient option for the provision of a city service or a maximizing strategy for air traffic flow or a solution to a model optimizing electric power demand, price, availability, and line capacities at a specific interval in time.

4. Design values guiding the policy formulation process embrace specificity and consistency over discretion, judgment, and trust.

When high social risks or critical objectives are at stake, we expect policy to avoid ambiguity, misunderstanding, or individual malfeasance. Policy formulators do not want to trust public objectives to the localized judgment of individuals in the implementation process. Under a design focused on the worst or best case, the possibilities of error by the arbitrary discretionary actions of individuals become magnified and intolerable. Trust and discretion seem self-evidently insufficient when designing policy to avoid the worst case.

These four policy design features are more and more evident in both public and private policy. Yet the assumptions are very problematic when it comes to high reliability management of our critical infrastructures. Consider them in light of the following three propositions culled from the preceding chapters and from findings in the wider literature.

Proposition I. There is an important divergence between dominant approaches to designing systems and the process of managing them.

The difference between design and management is above all a difference in cognitive orientation—ways of seeing, knowing, and responding—between different classes of professionals. Specifically, the macro-design orientation is very different from that of reliability professionals. These reliability professionals, as we have seen, are specially motivated to make things "turn out right," and they have largely unheralded skills for understanding the world that lies between the general principles and deductive orientations of designers and the case-by-case, experience-based preoccupation of many technicians and frontline operators.

Proposition II. Successful reliability management focuses less on safeguarding single-factor performance than on maintaining a set of key organizational processes within acceptable bandwidths.

The boundaries of these (*de jure* and *de facto*) bandwidths can be adjusted but only in proportion to improvements in the knowledge base (both formal and experiential) of the large technical system.

Proposition III. Despite the vulnerabilities they generate, centralization and interdependence among the component parts of a complex technical system can be significant managerial resources on behalf of high reliability.

What we see in the findings from the California case study is that both complexity and tightly coupled interactions can and often do serve as positive sources of high reliability performance, particularly in the just-in-time performance mode that puts a premium on adaptive equifinality.

If you compare the preceding four design assumptions with these three propositions, you can see how the world of macro-design—both policy and technical—has become more and more hostile to high reliability. Sadly, we see the same forces at work when it comes to technology as the purported key to reliability.

"ULTIMATELY, WE ARE LOOKING at more intelligent, automated, and decentralized tools to help control the reliability of the grid," said a DOE expert to us in 2001. Certainly, the reliability professionals we have interviewed at CAISO call for more and better tools and technology.

But here too, the associated risks are as enormous and real as any macro-design solution that promises to ensure reliable micro-operations directly. Rather than narrowing risk by replacing professionals with technology, new technologies can compel professionals to operate outside their domain under unstudied conditions, with few patterns and scenarios—thus in effect widening the task uncertainty of the professionals, not narrowing it. We saw and documented this the night we watched RTMA being introduced, which manifestly pulled professionals to the edge of their domain of competence.

We have no doubt that RTMA challenged the pattern recognition and scenario formulation skills of CAISO's reliability professionals. Why then would an organization committed to high reliability management such as CAISO undertake something as risky as RTMA?

One answer may well be organizational and political survival. A key informant told us that CAISO had bet its future on driving market and technology reforms, such as RTMA, because it was doubted that CAISO could long survive without them. While we do not doubt this to be a factor, we are asking a larger question here: Is there something in high reliability management that also pushes toward technological innovation? Clearly, high reliability managers are driven by the same constant search for improvement that was observed in the earlier high reliability organizations described in Chapter 4. Following are six factors that we believe explain how an organization committed to high reliability management sets into motion innovations that periodically challenge the skill base of its reliability professionals, *precisely because of this commitment to that management.*

1. For an organization pursuing high reliability management, the drive to reduce system volatility and increase options variety has to be central. It is implausible that such an organization would not want to ensure access to all four performance modes described in earlier chapters. Most assuredly it would be reluctant to be confined to one or two performance modes, particularly just-for-now. It is not surprising that proposals to reduce volatility and increase options are offered when reliability professionals themselves complain about high system volatility and few options. A gen dispatcher told us in 2007 that one of the great virtues of the proposed MRTU, for example, would be its guarantee of CAISO reserve requirements for real time.

2. In the case of CAISO, high reliability management centers on ensuring the real-time balancing of load and generation. Is it any surprise that technologies that offer real-time information and real-time controls appeal to CAISO? If the same technologies also enable levels of systemwide pattern recognition and contingency scenario formulation well beyond the capabilities of the operators themselves, one can readily understand the drive to search out and develop these initiatives. To conclude, as we do, that the technologies and software all too often do not fulfill their promise only argues that we be suspicious of their macro-design rhetoric, not of the search for improvement that gives rise to their appeal in the first place. Again, the reliability professionals we have interviewed are constantly talking about the need for new, better, and improved tools in the control room. At no point have they said to us that they are the only ones who can design those tools; at no point do we imply that only those proposals that operators originate can pass their reliability test.

3. We noted earlier that the retrospective approach of the formal design perspective—that reliability can be established on the basis of a past record of successful performances or deduced on the basis of prior principles—contrasts sharply with high reliability management. Indeed, it is the prospective orientation of control room operators that drives their management, in which, again, past success never guarantees future success. If reliability were equated only with recognizing past patterns, operators could never be reliable in the high reliability management sense. Their prospective orientation combined with their

constant search for improvement creates an opportunity space for new innovations to fill.

4. High reliability management is preeminently reliability-seeking behavior, and, as we have seen throughout, that means risk taking across multiple
types of risks (across different performance modes). Operators take risks—and
major risks at that—all the time as part of their management. That the reliability professionals are *managing* these multiple risks should not detract us
from recognizing that these are *major* and *multiple* risks being managed. Proposed software and technology innovations thus must be understood as at
times introducing risk as part of an overall strategy of managing risk.

5. As we saw earlier, high reliability and cost-effectiveness, far from being
mutually exclusive, can be mutually reinforcing. Yes, we have seen operators
requesting 103MW, knowing that they only need 100MW; but knowing that
they only need 100MW, they did not request 200MW. Requesting 200MW
"just to be on the safe side" would be tantamount to acting unprofessionally.
It would be wasteful, knowing what they know and with the high skills that
they have. Thus any innovation that enables operators to request less than
103MW to *reliably* ensure the 100MW to balance load and generation has appeal to them. As one operator described it, "Having more responsive steering
on a car allows you to maintain safety even when you come closer to the edge
[of the road]."

6. Last but not least, our discussion of the importance to high reliability
management of the working relationship between operators in the control
room and their wraparound support staff has direct bearing in understanding
the role of innovations that challenge operator skills. Even when operators are
successful at buffering output performance from wide variance in the inputs
and challenges they face, the wide variance remains. It thus is no surprise that
engineers and others in the wraparound are constantly searching for innovations to control that variance better. As we have described, the relationship is
more or less one of support by the wraparound for the control room, with less
support having a negative impact on operator management (Chapter 5).

More support, however, does not mean that operators and wraparound
staff have the same orientation. In fact, what we have been calling high reliability management can by this point be described as the result of the different
orientations of control room operators and wraparound staff (a phenomenon

described in detail in von Meier 1999). When we combine the design and protocol orientations of engineers with the micro-operations and pattern recognition skills of the operators we end up with the four positions of the reliability space. From the perspective of the reliability space, operators could hardly be reliability professionals tacking to the middle domain of unique knowledge without formal principles and design logics to work from and modify. At the same time, we saw that the biasing of the RTMA forecast by operators actually helped software designers improve their understanding of the system they were attempting to model.

If the wraparound staff and operators together map the entire cognitive space for high reliability, how could we not expect engineers in the wraparound to periodically come up with design innovations that challenge operators in the control room, *and* operators in the control room to continually challenge the skills of wraparound engineers to come up with better designs? We return to the role of improved operational designs as distinct from formal ones at the macro-level in Part Three.

In short, no one should take from this book the idea that high reliability management is inherently conservative, precautionary, or risk averse, in which few innovations that really challenge the skills of control operators and the wraparound staff are to be expected and in which any innovation that did challenge them would necessarily fail the reliability test. To repeat: high reliability management is its own form of risk appraisal and, as such, merits its own field of research.

CHAPTER II

MEASURING
HIGH RELIABILITY
MANAGEMENT

IT IS A PARADOX OF THE HIGH RELIABILITY management of complex infrastructures that its very success masks much of its inner workings. The bad news in the good news is that reliable critical infrastructures have insufficient failures to isolate the causal connections between specific causes and their positive or negative effects. This lack of information has implications for both management and measurement. In classic high reliability organizations, managers have worried about what factors exactly have been contributing to the reliability of their operations. As said before, they "run scared" in the face of this uncertainty, afraid to make significant changes in technology or organizational practices lest something upset successful performance. One engineer at a large electrical utility observed, "we may have overinvested in reliability, but who wants to make the cutbacks to find out?"

High reliability thus raises a major measurement challenge because of the complexity of managerial factors relative to the few instances of failure. This complexity presents the statistical problem of many variables and few cases. Yet careful measurement is instrumental to our understanding of high reliability management and its challenges. If we were to count only catastrophic failures, it would appear that almost no policy or technical changes (including electricity system restructuring in California) or even major internal organizational changes have much of an effect on reliability.[1] The lights stay on, so what's the problem? But if we are concerned about conditions that stress, disorient, or degrade the skills of reliability professionals, how can we identify these before a catastrophic failure?

The key to measuring high reliability management and its fluctuations lies in assessing risky situations that fall short of, but occur more frequently than, system failure. This concern with early warning parallels the focus of reliability professionals themselves who identify precursor variables—factors that while no direct threat to reliability could, through chains of indirect causation, lead to serious consequences. As seen in Chapter 5, this precursor zone of events or performance conditions is of great interest to high reliability managers as they attempt to gauge and gain sensitivity to its threshold.

Our efforts here regarding reliability measurement are directed toward (1) providing further evidence of the challenges to reliability management we found through our case description, (2) identifying quantitative measures or indicators of precursor zones with respect to high reliability performance in the CAISO control room, and (3) advancing our methods to monitor fluctuations in reliability conditions in a variety of infrastructures *prior to* catastrophic failure or collapse.

OUR INDICATORS RESEARCH follows directly from interviews with CAISO operators about what they mean when they say that some challenges push them to the edge of their abilities. We focus on the match between operator skills and tasks, or what we have called the control room's "reliability envelope." Given persisting operational challenges and changes at CAISO, indicators that can reveal edges of that envelope (that is, where task requirements and operator skills become unbalanced) are especially important in improving the early warning that can safeguard reliability.

We began by considering tasks and skills to be connected as challenges and responses are connected. Task requirements and operating challenges can be thought of as input variables and the response or performance variables can be identified as output variables. These inputs and outputs can be thought of as defining a reliability production function.

Through discussions with control room and other CAISO personnel, we identified input variables that potentially challenge the skills of operators. These included peak-load forecast error, hardware malfunctions (such as screens going blank), and software glitches. After further probing and discussion, we focused on one set of input-output relationships: the impact of task requirements imposed by the grid and software conditions of unscheduled outages, line mitigations (due to congestion), and RTMA software adjustment (load biasing) as input variables on CPS2 violations, the output variable.[2] Fortunately, each variable is measurable, and CAISO has archived the respective data over many years.

OUR INDICATORS ANALYSIS occurred in two phases. In phase one we used an initial database that ran from July 2004 through May 2006, inclusive, and gave us a maximum of 700 (daily) data points for the control room variables. Later, in phase two, we extended the database through the end of December 2006 and expanded our coverage to hourly data. The extension enlarged the database to over 900 days, with nearly 22,000 hourly data points for the major variables.[3]

In our framework for analyzing the skills of reliability managers, we have argued that these professionals promote reliability through their ability to deal with high input variance and reduce it to low output variance. The technique of linear regression analysis seems the ideal quantitative starting point for describing such a relationship and probing its implications. Regression allows a statistical analysis of how much the variance in a set of independent (input) variables accounts for variance observed in an independent (output) variable. The R^2 in a regression is a measure of the percentage of the output accounted for by the inputs. In addition the regression equation also measures the independent contribution each input variable makes to the final R^2.[4]

We undertook our regression analysis by mapping the multiple input variables (outages, mitigations, and high RTMA biasing) onto an individual output variable (CPS2 violations). The overall impact of input variables on output variance is measured by the tightness of fit, the R^2, between the regression equation and the actual data points associated with each variable. The coefficient associated with each input variable in the regression equation is a measure of the independent impact each is having on variance in the output, holding the other factors constant.

In our regression analysis, a low R^2 (assuming the relationship between our input variables and output variable is statistically significant) means that operators are controlling for the impact of input variables. The variance in the inputs is not being translated into instability of outputs. The challenges posed by the input variables are being well contained by the skills of the operators. Of course at the same time we do not know if other input variables might not be more closely associated with the remainder of the CPS2 variance not accounted for by our particular inputs. It is possible that another set of "edge" variables exists with respect to CPS2s. For example, transmission line outages and power import restrictions into California may account for the remaining variation in CPS2s.

A high R^2, in contrast, suggests a heightened sensitivity of grid management performance in the output variable with respect to fluctuations in the input variables, indicating that grid managers and operators are nearing a performance edge regarding their ability to contain the effects of the specific input conditions. A high R^2 signifies two features of cognitive concern for operators. First, the input variables are becoming more difficult to control when it comes to limiting their impact on output variables, such as CPS2 violations. Second, the heightened impact of the specific input variables on this output variable can mean that operators are less able to attend to other variables and edges that may well be important during these same periods of stress. If operators are focused primarily on one edge for a prolonged period of time to the exclusion of attending to other potential edges, the resilience of the control room to rebound from shocks as well as to anticipate those shocks could well be reduced.

Following are summarized findings of our first and second phases of research undertaken between November 2006 and April 2007. All pertain to the one particular performance edge—that connecting RTMA biasing, outages, and mitigations to CPS2 performance. As no high reliability manager we know focuses on one and only one edge all the time, the results are illustrative of only one set of challenges to reliability and methods to identify and evaluate them.

THE EARLIER RESEARCH discussed in Chapter 6 identified a period when CAISO control room operators reported they were being pushed closer to the edge of their reliability envelope in terms of CPS2 violations. We can now confirm this statistically. Soon after the introduction of the RTMA dispatching system, unscheduled outages, line congestion mitigations, and the need to bias the RTMA software itself had a discontinuously greater impact on CPS2 violations than previously or indeed later. Starting when RTMA bias data initially became available—November 11[5]—to December 31, 2004, our statistical findings reveal the problems dramatically.

Table 11.1 sets the baseline for comparing results. It shows the relationship between high RTMA bias, unscheduled outages, and total mitigations on one hand and CPS2 on the other over the first phase period for which we have data, November 2004–May 2006.

The model as a whole, as well as the coefficients of its three input variables, is statistically significant. Contrast the Table 11.1 baseline with its adjusted R^2 of roughly 0.23, to Table 11.2, with the much higher R^2 for the same variables during the period just after the introduction of RTMA. This was when operators explicitly reported significant difficulty in reliability management, and indeed it was difficult. The variance explained increases to 46 percent, and all coefficients, along with the F-statistic, remain statistically significant as well.

That the variance more than doubled illustrates a significant decrease in the ability of operators to buffer CPS2s from the impact of the input factors.[6] The skills that operators normally use to manage unexpected outages and increased mitigations so that they do not worsen CPS2 violations became much more difficult to exercise or were less effective during the startup period in which RTMA was introduced. In our interviews with operators earlier they had complained about the difficulties of using RTMA, both in having to debug its programming and in paying close attention to its dispatches to make sure they were not wildly erroneous.

To what extent is the output variable (CPS2s) sensitive to *each* input variable during an edge subperiod? One way to illustrate the relative importance of individual input variables is to compare their regression coefficients during

TABLE 11.1.

Summary output for high RTMA bias, unscheduled outages, and total mitigations against CPS2 violations, November 11, 2004–May 2006

Regression Statistics

Multiple R	0.478669409
R²	0.229124403
Adjusted R²	**0.225016718**
Standard Error	4.20608558
Observations	567

ANOVA

	df	SS	MS	F	Significance F
Regression	3	2960.414175	986.8047249	55.77944501	1.40104E-31
Residual	563	9960.139617	17.69118937		
Total	566	12920.55379			

	Coefficients	Standard Error	t Stat	P-value	Lower 95%	Upper 95%
Intercept	1.400652638	0.581155674	2.39364042	0.01700843	0.251297796	2.550007479
High RTMA Bias	0.031794547	0.008049907	3.949678913	8.81879E-05	0.015983029	0.047606065
Unscheduled Outages	0.094896986	0.010859559	8.738567241	2.70838E-17	0.073566788	0.116227185
Total Mitigations	0.58376443	0.07890819	7.39563789	5.13381E-13	0.42858635	0.73856852

TABLE 11.2.

Summary output for high RTMA bias, unscheduled outages, and total mitigations against CPS2 violations, November 11–December 2004

Regression Statistics

Multiple R	0.702038137
R²	0.492857546
Adjusted R²	**0.460486751**
Standard Error	3.556683693
Observations	51

ANOVA

	df	SS	MS	F	Significance F
Regression	3	577.8029934	192.6009978	15.22537666	4.68374E-07
Residual	47	594.5499478	12.64999889		
Total	50	1172.352941			

	Coefficients	Standard Error	t Stat	P-value	Lower 95%	Upper 95%
Intercept	-0.048563722	1.395604418	-0.034797627	0.972388482	-2.856157625	2.75903018
High RTMA Bias	0.055635045	0.018780065	2.962452272	0.004775903	0.017854429	0.093415662
Unscheduled Outages	0.092521121	0.041441223	2.232586663	0.030378399	0.009152136	0.175890107
Total Mitigations	0.675037138	0.23256758	2.902564604	0.005619184	0.207175188	1.142899087

periods of stress to the respective coefficients during the baseline period. The regression coefficient estimates the change in the output variable due to a unit change in the input variable, holding other input variables constant. For example, going back to the Table 11.1 baseline for the CPS2 edge, we see that the regression coefficient for the input variable high RTMA bias is approximately +0.032. This means that for a one-unit increase in the number of time periods of high RTMA bias, you would expect a +0.032-unit increase in number of CPS2 violations, holding unscheduled outages and total mitigations constant. During the RTMA introductory subperiod, CPS2 was found to be clearly more sensitive than in the baseline period to changes in high RTMA bias and mitigations. In each case, a unit change in bias and mitigations led to a greater unit change in CPS2 than in the baseline period (coefficients of 0.056 and 0.675 for subperiod bias and mitigations [Table 11.2] compared to 0.032 and 0.584 for respective baseline values [Table 11.1]).

Being closer to an edge can also mean the input variables become more sensitive *to each other* as they become more important in determining changes in the output variable. For instance, increases in outages may require more line mitigations in order to reroute power from alternative locations. One way to check for mutual sensitivity is to look at the cross-correlations of the input variables during the stress periods to determine if they change from baseline correlations. Table 11.3 shows the respective correlation matrices for the baseline period and RTMA introduction with respect to the CPS2 edge.

The correlations between outages and mitigations increased during the RTMA startup period, as did the correlation between the three independent variables and the output variable CPS2. (Note that the weak cross-correlations in the baseline period confirm the operators' view that these challenges are largely independent factors during "normal" periods.)

In brief, the findings indicate that the mutual sensitivity among independent variables increased along with their relative importance during the RTMA introduction period. This result, along with the substantially higher-than-baseline R^2 during this subperiod, foretells the kinds of stresses to expect after the introduction of any new major piece of market software at CAISO. To put it in the more formal terms of Chapter 7, when variables determined by prior analysis to be largely independent of each other end up interacting during periods of heightened stress in heretofore unseen ways, we can expect rather

TABLE 11.3.

Correlation matrix for the high RTMA bias, unscheduled outages, and total mitigations against CPS2 violations

A. Baseline (November 11, 2004–May 31, 2006)

	RTMA Bias	Outages	Mitigations	CPS2 Violations
High RTMA Bias	1			
Unscheduled Outages	0.16739785	1		
Total Mitigations	0.12161903	0.134049472	1	
CPS2 Violations	0.128317839	0.344503392	0.34195265I	1

B. RTMA Introduction (November 11–December 31, 2004)

	RTMA Bias	Outages	Mitigations	CPS2 Violations
High RTMA Bias	1			
Unscheduled Outages	0.087177093	1		
Total Mitigations	0.296390299	0.50164651I	1	
CPS2 Violations	0.454588648	0.479874825	0.595176227	1

more bandwidth management under active analysis to be taking place in the control room than bandwidth management under prior, anticipatory analysis.

THE SECOND PHASE of our indicators study enabled us to better identify and measure movements to and away from the performance edges and the resilience involved, using both daily and hourly data over an extended database. In the second phase, we identified additional movements to and away from the CPS2 performance edge over the expanded study period from July 2004 through December 2006 inclusive (over nine hundred days). To do this, we developed a methodology that enables the tracking of the movements over time.

We also developed a typology of good and bad periods with respect to control room management of variable inputs—high RTMA biasing, unscheduled outages, and total mitigations—for stable outputs. We now are able to differentiate edges that are difficult because the tasks are *intensifying* for operators versus periods in which their skills to meet these tasks are being *degraded* in some fashion.

These methods and analysis enable us to learn more about the impact of specific input variables, particularly high RTMA software bias. We now know that while the number of times control room operators use high bias has gone down (a positive finding), when they do have to resort to a high bias, the consequences have a generally increasing and negative impact on CPS2 violations. We also can see more specifically the disorienting effect that the introduction of RTMA had on control room performance.

Let us examine each finding briefly.

OPERATORS MOVE CLOSE to a performance edge not only because they are pushed there, but also because doing so at times has benefits, not just cost. Moving closer to the CPS2 edge may reflect a way operators allocate their allowable CPS2 violations.[7] There are occasions when CPS2s become a resource for achieving reliability, for instance, when a CPS2 violation is incurred to avoid a worse situation or as a system probe that adds new scenarios to the operators' repertoire and enlarges their inventory of recognized system patterns.

To measure how quickly operators move to and away from an edge, we developed the graphic display of what we term an "edge resilience trajectory" (ERT). An ERT shows the relationship between reliability and resilience in the control room in three ways. One measure of the ERT is to track a moving range of edge R^2s across the baseline period. Taking essentially a fifty-day period (for statistical significance), the range can be moved forward day by day to generate a series of R^2s for the CPS2 edge across the study period of July 1, 2004, through December 31, 2006.[8] In this way, the measurement of movement toward or away from an edge takes account of what happens in fifty-day increments. We can see how resilience operates over the shorter term of some two months.

A second element in an ERT is to generate a series of R^2s by adding one day at a time to an initial subperiod, such that the final R^2 becomes the baseline R^2 for the entire period. In this way, the final measurement is a cumulative one that takes into account what happened at the beginning of the study period. This add-on approach affords a wider-angle lens that gives a long-term look at how operations return to baseline after a shock, excluding short-term fluctuations in input sensitivity.

The third part of an ERT marks the days when the actual number of CPS2 violations exceeded one positive standard deviation away from their baseline average. These are days when operators were noticeably less able to control the output variable, CPS2 violations, in relation to high RTMA bias, outages, and mitigations. The postulate here is that because high reliability management is about controlling large input variance to produce low output variance and stable performance, the days when output is neither low nor stable have been particularly challenging to grid managers.

In sum, an ERT reveals a moving range of edge R^2s (a short-term way to see resilience), the cumulative response of operators to those movements (a way to see the longer-term effect of resilience), and the days when control room operations are not resilient enough to prevent additional risk to high reliability management with respect to an edge.

Figure 11.1 shows the edge resilience trajectory for CPS2 output variable for the expanded study period. The light triangles indicate the moving range R^2s, the light circles indicate the cumulative add-on R^2s (the gaps are where the R^2s did not have significant F-statistics), and the dark triangles or circles indicate days on which actual CPS2 violations substantially differ from baseline values (the add-on curve is used when no significant R^2s are available in the moving range).

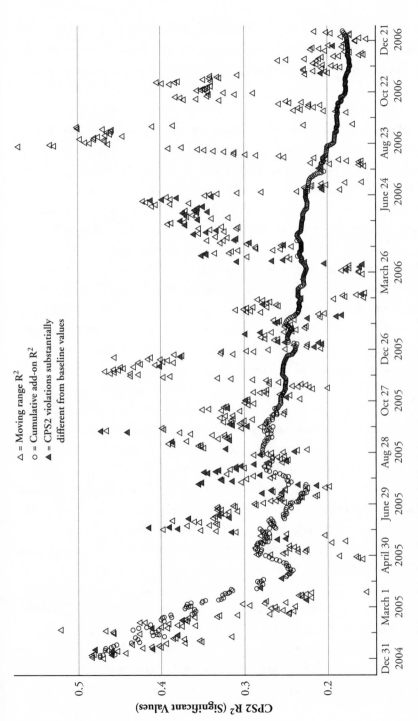

Δ = Moving range R²
o = Cumulative add-on R²
▲ = CPS2 violations substantially
different from baseline values

CPS2 R² (Significant Values)

FIGURE 11.1. Edge resilience trajectory for CPS2, November 11, 2004–December 31, 2006

When we are now asked, "What is resilience?" we point to Figure 11.1 and say, "That, in CAISO's grid management, is resilience." This is what resilience is about: moving to and away from edges in ways that steadily—though rarely smoothly—improve performance over time. We see this improvement in five ways.

First, the ERTs provide independent confirmation of the periods of stress we were told about by control operators. The spikes in the moving range around the RTMA introduction and other periods mentioned to us (for example, the C1 network changeover mentioned in endnote 6) are distinct and visible. The periods also include a number of days in which CPS2 values significantly exceeded the baseline. It appears that more higher-than-baseline CPS2 days are associated with higher rather than lower R^2s in the moving range—again indicating problems associated with buffering input variables.

Second, other spikes indicate periods of stress about which we were informed in our interviews, but were unable otherwise to identify. We were told in the first phase of our indicator study that the "shoulder" (seasonal change) months of April–May and September–October could be important stressors because it becomes harder to predict peak loads and the time of their occurrence. The CPS2 ERT confirms spikes around the shoulder months of 2005 and April–May 2006. The frequency in these months of higher-than-baseline CPS2 violations (the dark values) also supports the finding.

Third, several spikes identify possible periods of stress of which we have yet to be informed but which merit closer examination. The spike between March and June 2006 seems to involve more than the shoulder months, and the frequency of days in which CPS2 problems were significantly higher than baseline is quite visible here.

Fourth, an ERT distinguishes spikes from each other. The spike of late July–September 2006 is dramatic, but with virtually no days of higher-than-baseline CPS2s. Compare this spike to the one of roughly a year earlier, July–October 2005, with its many higher-than-baseline CPS2 violations. The difference between the two spikes can be summarized in a phrase: Resource Adequacy. The earlier spike occurred before the California Public Utilities Commission adopted its Resource Adequacy rule requiring distribution utilities to acquire 115 to 117 percent of generating capacity over forecast peak demand. The later spike occurred after its adoption. Look carefully at Figure 11.1 and you see that July 21, 2006, the day when CAISO successfully met an unprecedented peak load (Chapter 6), is at the base of the late-July–September 2006 spike. Notably, CPS2 levels are not increasing in the spike, albeit a larger proportion of CPS2s are attributable to the input variables.

Fifth, the overall trend in the ERTs is downward and improving; after a shock like RTMA, operators bounce back to normal operations, though "normal" has now strategically changed. The short-term spikes may well be helping to promote learning that drives the trajectory generally downward, with fewer days of CPS2 being above baseline. Operators who are challenged by difficult situations respond by making corrections through formulating new scenarios and recognizing new patterns that improve cumulative performance. Again, operators take risks to reduce risks. Note that the corrections take relatively few days; the figure shows fairly steep spikes before an inflection point is reached and then input sensitivity quickly resumes a downward direction.

That said, being at a spike and with higher-than-baseline CPS2 values involves risks (again that task requirements threaten to exceed the level of operator skill) even if the periods generate scenarios and patterns that improve system reliability. The high spike in August–September 2006 and the smaller one afterward have had a statistical impact by pulling up the hitherto declining cumulative add-on ERT.

Finally, as Figure 11.1 shows, even for *low-R^2 periods,* there are days when the actual values of CPS2 are well above their baseline levels, once again confirming what operators have told us: other factors aside from large numbers of unscheduled outages, high RTMA bias, and mitigations can and do challenge reliability management.

Edge resilience trajectories can be generated for specific subperiods of stress. Figure 11.2 shows the CPS2 ERT for the period after the RTMA introduction. Again, the triangles indicate the moving range R^2s and the dark circles the cumulative add-on R^2s, where gaps in the plot indicate the calculated R^2s did not have significant F-statistics. (Days of higher-than-baseline CPS2 problems have not been identified for clarity of presentation.)

The downward trajectory of the edge is clear, as operators became better at reducing periods of high biasing and controlling for impacts of outages and

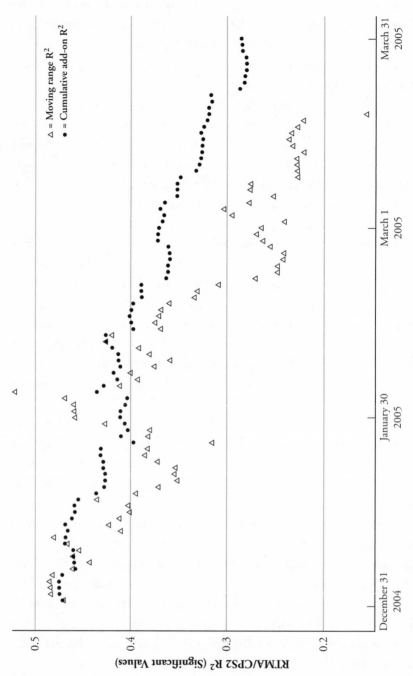

FIGURE 11.2. Edge resilience trajectory for CPS2 during RTMA introduction

mitigations between November 2004 and the end of March 2005. Improvements continued, though later periods had their own shocks, for example, the C1 network changeover that followed.

An edge resilience trajectory, like the one in Figure 11.2, reminds managers that major changes entail days when it is difficult to manage in a highly reliable fashion. Second, it can serve as a template that managers use to set or monitor management goals, such as establishing advanced training and simulator work with surprises and unexpected scenarios for operators so that they can learn to handle a new shock better and rebound more quickly. Third, the ERT can help identify when operators are not getting ahead of conditions induced by a major change and thus are not improving as well as they have in the past with respect to a given performance edge.

Judging performance by ERTs must be tempered by caution. The assumption is that the control room is better off once the shock is absorbed and operators have rebounded to the new normal. The risks facing operators are reduced, but still risk remains. Reducing the frequency of high RTMA biasing has had both a positive and negative side. Ongoing reduction in the number of times control room operators have to use high bias is positive. However, when they actually do bias highly, the consequences with respect to CPS2 violations seem to be increasing. Figure 11.3 shows that the frequency of high biasing has gone down considerably, but the impact of high biasing when it does occur, as measured in the bias regression coefficient, has trended upward over the study period.

OUR DEFINITION AND MEASUREMENT of high reliability management—managing wide input variance to achieve low output variance—allows us to better distinguish the types of challenges control room operators face when *actually at the edge of their performance.* We do this by comparing the means and standard deviations of the input variables and output variables for each moving (~50-day) range of R^2s to the respective baseline values (means and standard deviations) over the entire study period. Whether the R^2s of the moving range are high or low and whether the range means and standard deviations of specific variables are above or below the respective baseline values tells a great deal about the challenges facing operators as they move to and away from their performance edges.

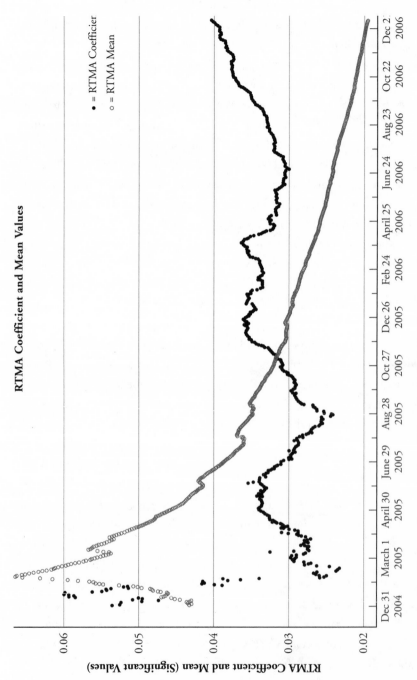

FIGURE II.3. Comparison of means and coefficients for high RTMA biasing

A typology of challenges for the CPS2 edge is developed through several steps. The first cut is to determine if the moving range subperiod has a low or a high R^2. (In the following analysis, the R^2s were ranked from highest to lowest, divided into three equal cohorts with the middle one excluded for clarity of analysis.) Again, a high R^2 indicates that movements in the inputs account for more movement in the output than would be the case if the R^2 were low. High R^2s indicate moving to or remaining at an edge, while low R^2s indicate operators are away or moving away from that edge.

Once the highest and lowest R^2s are grouped separately, the second cut is to compare input and output values over the range to their respective baseline values. Four sets of mutually exclusive conditions are possible in each R^2 grouping, depending on whether or not subperiod input and output values exceed baseline values. Each cell represents possible conditions facing the control room, which would pose distinct challenges to operators when the R^2 is high. By determining how many subperiods actually fall within each cell and when they occur (and match to the Figure 11.1 ERT), we can better understand the actual types of conditions facing operators and their responses.

By and large, periods of *low R^2s* reflect more favorable conditions than do periods of high R^2s. In the former, whether input and output means and standard deviations are above or below their baseline values, operators continue to buffer CPS2 violations from the effect of fluctuations in high biasing, unscheduled outages, and total mitigations during the subperiods. That said, conditions vary in importance depending on whether subperiod means and standard deviations for edge variables exceed or fall below baseline values:

- If means and dispersion values associated with CPS2 violations are lower than their baseline values, but the means and dispersion values are greater-than-baseline for the input variables of high biasing, unscheduled outages, and total mitigations, then we would say operators are being especially effective in buffering movements in CPS2 violations from movements in that edge's input variables. Output variance is kept low even in the face of high input variance.

- When the input values are themselves at or below their baseline values, along with the output values being lower-than-baseline, then management conditions with respect to this edge are ideal for operators to pay attention to something more important. Because effective buffering is going on, any improvements in conditions surrounding high biasing, unscheduled outages, and mitigations free up time and resources for operators to focus on other priorities.

• If, however, output values are higher-than-baseline in these subperiods of effective buffering with respect to this edge, then it would be because something else is affecting the output variable in negative ways above and beyond the input variables being managed in this edge.

  ○ If CPS2 violations are higher than usual but input means and standard deviations are lower than usual, then another edge as-yet-unidentified might be at work in accounting for the increase in CPS2 violations.

  ○ Alternatively, if CPS2 violations and the input values of high biasing, unscheduled outages, and total mitigations are all higher than their baseline values, then something else is intervening between these inputs and that output to account for the low R^2s. For example, something may be happening in or to the grid (such as an extreme heatwave) that is worsening the CPS2 and all the input variables, such as mitigations and outages, but not connecting them more closely to each other.

High reliability managers would hope these last conditions are few and far between, because on days in which they are present a widespread set of difficult conditions must be present. Fortunately, as Table 11.4A indicates, we found no periods with such conditions.

Periods of *high R^2s,* in contrast, pose unfavorable conditions, and in some cases, severe and distinct challenges to high reliability managers (Table 11.4B). In these conditions, tasks may intensify, thereby increasing difficulties with respect to the edge of concern. It is even possible for skills to degrade in the face of an unchanging set of edge tasks, indicated by periods in which inputs have stable or decreasing values. Let us review the possibilities separately:

• Operators face task intensification whenever input values for a subperiod are above their respective baseline values (more frequent periods of RTMA biasing or more generator outages, for instance) during those times when operators find it more difficult to moderate the effects of these variables on CPS2s.

TABLE 11.4.

Actual distribution of moving range R2s across operating conditions and challenges—CPS2 edge

A. *Challenges under Low R^2 Conditions*

		Output Means and Standard Deviations	
		Over baseline output values	*At or below baseline output values*
Input Means and Standard Deviations	*Over baseline input values*	Indication of a decline in a managed connection between input and output variables [o]*	Indication of ability to buffer high input variance to achieve low output variance [39]
	At or below baseline input values	Indication of presence of other edge(s) having an impact [o]	Indication of taking advantage of easier task environment [5]

B. *Challenges under High R^2 Conditions*

		Output Means and Standard Deviations	
		Over baseline output values	*At or below baseline output values*
Input Means and Standard Deviations	*Over baseline input values—task intensification*	Indication of specific threat to operator management [o]	Indication of specific ability to achieve low output variance during heightened sensitivity to input variables [5]
	At or below baseline input values	Indication of possible loss of skill base in operator management [10]	Indication that improvement in inputs during heightened sensitivity is leading improvement in outputs [14]

NOTE: *Number of subperiod ranges in brackets.

o   If at the same time that task intensification is going on, output values are at or lower than baseline, we would say that operators are coping in a reliable fashion. They are managing wide input variance while still controlling output variance under already difficult conditions with respect to this edge.

- If, however, at the same time that task intensification is going on, output values are higher than baseline, then conditions exist that threaten high reliability management. Here operators find it difficult to do what they normally do well. Conditions on the input side are worse than usual, output values have also worsened, and these outputs are more heavily tied to the input conditions than during a baseline period of reliability. Operators are at the edge of their reliability skills, if not being pushed over them.

- It is also possible for input values to be at or below their baseline values for some of the subperiods during these difficult times at the edge of performance. Two possible conditions arise, one worrisome, the other reassuring:

  - The means and standard deviations associated with high biasing, unscheduled outages, and total mitigations could be lower than baseline values, even while the mean and standard deviation associated with the output variable are higher in this subperiod. CPS2 values are going up under already difficult conditions and yet input values are going down, when compared to their baseline. This troubling case suggests to us that the skills of operators concerning these inputs are being threatened in ways that have little to do with their tasks intensifying. The number and dispersion of high bias intervals are declining, as are the values in unscheduled outages and mitigations, but conditions are still difficult and getting worse with respect to this edge. It is as if the operator skills, relative to these variables, are not as effective as before.

  - Finally, there could be a subperiod during difficult times, during which both input and output values are going down at the same time. When this happens, it clearly seems to be the case that improvement in input conditions is directly accounting for improvements in the output conditions.

Table 11.4 sets out the typology and the distribution of subperiod ranges in the CPS2 edge across the cells for the study period November 11, 2004, through December 31, 2006.[9]

From a reliability perspective, the distinctions in the high R^2 cohort are the most pressing. The good news is that no sustained threats to high reliability management (when both input and output values are increasing) were found when operators were at the edge of their CPS2 performance (when the sub-period R^2s were high). Equally good news are the five subperiods starting in the second week of September and finishing at the end of October and beginning of November 2006, when operators managed in a highly reliable fashion while at their CPS2 edge (Figure 11.1 also shows operators moving away from the edge at that time).

The bad news is that ten subperiods were recorded, starting and clustered around early December 2004 and ending around late January and the beginning of February 2005, during which operator skills were challenged while at the edge of their reliability envelope. It is noteworthy that the ten subperiods during which operator skills were under greatest challenge during the study period occurred more or less in sequence just after the introduction of RTMA during the bounce-back period identified in Figure 11.2. Clearly the introduction of RTMA disoriented control room operators in ways that upset their ability to be as skillful as they normally are. The skills for scenario formulation and pattern recognition were significantly threatened at that time.

If we turn to the low R^2 cohort (Table 11.4A), it is reassuring that most of the thirty-nine subperiods during which operators were managing wide input variance to produce low and stable output variance were fairly recent. Here too there was a great deal of clustering of the subperiods ending in November and December 2006. Also as we hoped, there were no subperiods in which operator management of CPS2 violations in response to high biasing, unscheduled outages, and mitigations was weakened by intervening factors or edges not yet mentioned to us by control room operators.

Finally, it is useful to triangulate on these findings. In particular, do the high R^2s subperiods of good management (the right side of Table 11.4B) differ significantly, in statistical terms, from those low R^2 subperiods (the right side of Table 11.4A)? Where subsample sizes permit, a conventional way to test for any independent effect a subperiod has is to use a dummy variable that treats the subperiod itself as an independent input variable in the regression equation. Treating the subperiod as a variable lets us determine if the subperiod coefficient is statistically significant. If so, the independent effect of this subperiod on CPS2 violations is significant overall. We tested this by analyzing four input variables (high RTMA bias, outages, mitigations, and the dummy variable set at 0 for low R^2 and 1 for high R^2 subperiods) against CPS2 violations in a regression equation. We found the F-statistic and all four regression coefficients to be statistically significant at the 0.05 level or better. Moreover, the dummy variable coefficient was positive, confirming that being in the high R^2 set of days presents more difficulties with respect to managing for CPS2 violations than being in the low set, other things being equal. Thus we can be fairly confident that the two sets of days are distinct when it comes to managing CPS2 violations.[10]

Do such differences survive when the comparison is on an hourly basis? To gain a finer-grained picture of what good and bad days look like in the control room, it is useful to turn to an hourly analysis of spike periods identified in Figure 11.1 and good and bad periods identified in the Table 11.4 typology.

STARTING WITH MORE RECENT SPIKES in Figure 11.1, we see that the period at the end of April and beginning May 2006 was especially troublesome to operators in terms of controlling for CPS2 violations at the CPS2 edge (note the density of dark triangles during this subperiod). Figure 11.4 compares the curve for hourly mean CPS2 violations over the entire study period (July 2004–December 2006) with its counterpart curve for the subperiod spike April 25–May 24, 2006.[11]

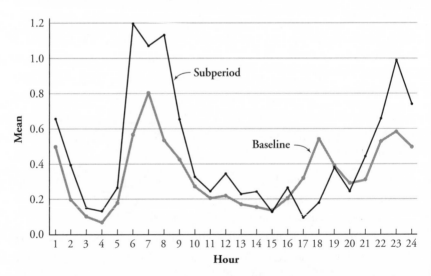

FIGURE 11.4. Comparison of subperiod and baseline CPS2 means, April 25–May 24, 2006

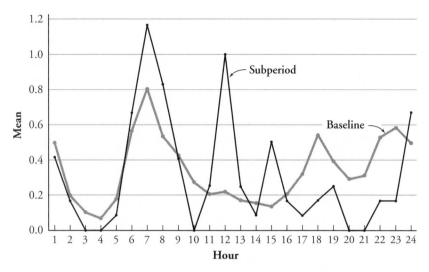

FIGURE 11.5. Comparison of subperiod and baseline CPS2 means, April 25–May 5, 2005

CPS2 violations were generally higher than baseline during the subperiod spike and considerably higher for the more problematic Hour-Ending (HE)7 (the seventh hour of operation in a day) and HE23. HE6 and HE8 were also difficult during the spike, while HE18 appears to have been less troublesome than usual. Comparing spike and baseline standard deviations virtually matches the profiles in Figure 11.4. In short, this mid-2006 spike was considerably more challenging on an hourly basis for the high reliability objective of ensuring low output variance and stable performance.

Returning to Figure 11.1 and starting now from the left after the RTMA introduction, several subperiods of high stress (that is, spikes of higher R^2s having higher than average CPS2 violations) are clearly visible between the ends of April and October 2005. Figure 11.5 compares the hour-by-hour CPS2 violations for the first spike (roughly eleven days between April 25 and May 5, 2005) to the baseline hourly curve for entire study period.

The two curves clearly do not converge and differ in interesting ways when compared to the subperiod in Figure 11.4. HE7 is more difficult than usual in this ten-day 2005 subperiod, as are HE12 and HE15 in terms of controlling for CPS2 violations. Also, HE24 is more difficult while HE16 and HE23 appear to be less troublesome than usual. Do we see other evidence of "off-peak" hourly peaks in CPS2 violations? If we define bad days as being days with fourteen or more CPS2 violations (the NERC reliability criterion), then the answer is yes. Of the 914 days in the expanded dataset, 97 days recorded more than fourteen CPS2 violations. Figure 11.6 compares the hourly averages over these days to the hourly mean curve for those days with fourteen or fewer violations and to the baseline curve.

Because less than 11 percent of the dataset were bad days in terms of CPS2 violations, the baseline and under-fourteen mean hourly curves virtually coincide. Note, however, the irregular peaking during HE20 during bad days compared to the typical dipping in violations at the same time in the baseline and good-day curves. Such findings strongly suggest the need for real-time tracking of CPS2 violations to determine if *any* off-peak hours in terms of violations are diverging from what is normally observed by way of pattern recognition and scenario formulation. In our framework, off-peak hourly spikes in a major output variable indicate that system volatility is changing in important ways.

The other way of defining good and bad days is by our typology. The ten subperiods of ~50 days in Table 11.4B, when possible loss of operator skill base was observed, can be thought of as bad days for high reliability management at the CPS2 edge. The five subperiods in Table 11.4B, when the ability of opera-

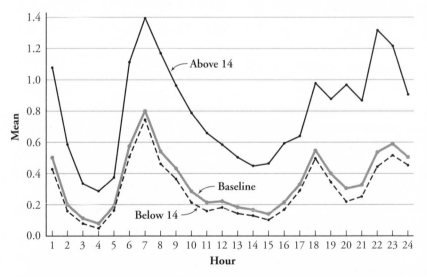

FIGURE 11.6. Comparison of baseline above fourteen and equal or below fourteen CPS2 violations means

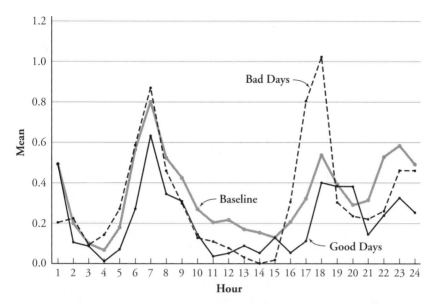

FIGURE II.7. Comparison of baseline, good and bad days subperiods CPS2 means

tors to achieve low output variance during the heightened sensitivity to the input variables was observed, can be thought of as good days for high reliability management at the edge. Are there differences between good and bad subperiods in terms of their output variables profiled hour by hour? Figure 11.7 compares the subperiod profiles against the baseline hourly CPS2 profile.

As the typology suggests, CPS2 violations are higher in the bad subperiods than in the good ones, with the baseline hourly profile somewhere in between during HE7 and HE18. That said, overall the three CPS2 hourly curves track each other very closely.

However, the profiles for the high RTMA biasing variable differ dramatically between good and bad subperiods (Figure 11.8).

The disorientating effect that the introduction of immature software for RTMA had on the skill base of operators is patent. The curve for high RTMA biasing during the ten subperiods characterized by operator disorientation and possible skill deterioration after the introduction of RTMA contrasts sharply with the lower curve in Figure 11.8, whereas nearly two years later operators demonstrated the ability to achieve lower and more stable CPS2 violations even during periods in which outages, mitigations, and RTMA bias were of major concern. Improvements in reliability, in other words, can be realized

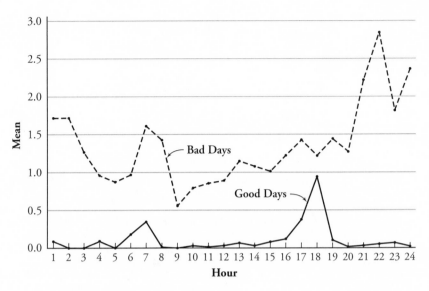

FIGURE 11.8. Comparison of good and bad day subperiods in terms of high RTMA bias means

even when operators are at or near an edge, when they learn to manage the challenges they face more effectively.

BY THIS POINT, we hope it is clear that researchers do have the means to illustrate and confirm empirically what they are observing in the way of high reliability management in the control rooms of society's critical infrastructures. Challenges to high reliability management can be assessed and evaluated without waiting for catastrophes to occur. There are answers now to the earlier question: If the lights stay on, why should we care? We care because of the spikes in performance challenges we saw in Figure 11.1 and the specific response problems of concern in the Table 11.4 typology.

We also care because those factors that push and pull operators to their edge are not reflected in any price movements in the electricity sector. Price fluctuations in electricity rarely if ever match the challenges operators face in providing that electricity reliably. Yet some of these challenges, such as generator outages and the need to make complex line mitigations, are closely connected to seller and consumer behavior.

Although our indicators do not allow calculation of specific probabilities of system failure posed by any edge period, an organization such as CAISO is now in a position to identify with some precision when it has entered into a precursor zone with respect to reliability conditions. As one former CAISO control room official put it, the indicators can "tell us if today is better or worse than yesterday, and that is a major advance." In particular, CAISO can now calculate the associated costs (once indicators have identified the length, nature, and severity) of an edge period. CPS2 violations can entail fines, higher than average RTMA bias imposes costs, while worsening mitigations and unscheduled outages pose their own expenses to CAISO and its customers. The indicators provide data from which to compute and differentiate these "edge" costs. We return to the importance that the more careful measurement of high reliability management has for the full-cost pricing of reliability in Part Three.

The starting point, however, for any such research in our critical infrastructures must be the close and careful observation of reliability professionals. Although we hope the concepts in Parts One and Two help focus those observations, and that the analysis in this chapter can assist other researchers in their empirical work, nothing here is intended to substitute for the hard work of interviews, on-site observations, and discussions researchers must undertake to truly understand the critical infrastructures of interest to them.

THE CHALLENGE OF HIGH RELIABILITY
Implications

WHAT DO THE CASE STUDY AND CONCEPTS developed in Parts One and Two imply for the high reliability management of other infrastructures and other organizational settings? A great deal, we argue in the concluding chapters.

The temptation is to jump in and detail specific lessons for specific infrastructures. In this fashion, we would generalize from our case and the supporting literature by offering recommendations to all manner of managers, regulators, policymakers, and other stakeholders in other critical infrastructures. We are loath to do so. To do so would place us in the very position of macro-design about which we have cautioned.

Instead, we believe that a reader already informed about other critical infrastructures is in a more sensible position to draw his or her own implications. Further, we find it difficult to forecast what is specifically ahead for

CAISO, let alone for other critical infrastructures we know less well, given the rate of organizational change we have observed over the years.

The literature on what we have been calling reliability professionals and their role in managing other critical infrastructures is scant. It is important, however, to note that what literature there is confirms both the importance of reliability professionals and how their composition varies from infrastructure to infrastructure, and from time to time within a given infrastructure.

WE CAN START WITH EARLIER WORK on other critical infrastructures to which we have contributed and from which we have drawn insights for *High Reliability Management.* Van Eeten and Roe (2002) describe the control rooms of large water supplies they visited as part of their study of major ecosystem restoration projects in Washington, Florida, and California:

These rooms are the one place in our case studies where we will find the technical competence, complex activities, high performance at peak levels, search for improvements, teamwork, pressures for safety, multiple sources of information and cross-checks, and the best example of the culture of reliability, all working through technologies and systems that build in sophisticated redundancies to buffer against potential failure and catastrophe. . . . The control room in these systems is the decision space for high reliability management (109).

Just what were line operators doing by way of management? Van Eeten and Roe describe their introduction to the Florida control room:

Steve takes us down to the control room of the water management district. Tommie shows us around. Over there are the workstations. We see some of the many spreadsheets. That side, a new intelligent warning system station is being debugged by engineers before being passed on to control-room line operators. The line technicians want to know for sure what they are seeing is real, real-time information and not bugs in the software. Tommie says they are hiring an environmental sciences person to manage wetlands in real time for stormwater treatment. He will work with the technicians and engineers in the control room (238–239).

A great deal of what actually occurred in the control rooms of the water supplies matches what we have observed in the CAISO control room. As van Eeten and Roe describe,

Line operators in the control room focus primarily on keeping the system stable within specified bandwidths. The bandwidths include requirements regarding not only water supply, but also water quality, flood control, power generation, and other activities. . . . Outside the control room, these bandwidths are negotiated among the agencies involved (152).

These quotes point to a difference with respect to reliability professionals between those in the CAISO control room and those studied in the water supply systems. Although environmental issues are important to CAISO, no environmental specialist has been stationed as a line operator in its control room yet. Further, although CAISO engineers are an extremely significant part of its wraparound support staff, they have yet to be permanently stationed in the control room proper at the time of writing, unlike what can be found in some grid control rooms elsewhere.

Last, van Eeten and Roe found that the three water supplies they studied were effectively managed as an interagency effort involving not just the water supply agencies but agencies charged to ensure other reliability mandates associated with the real-time provision of water, including those related to fish, game, habitat, and endangered species (van Eeten and Roe 2002, 152–153).[1] This interagency expansion of a hands-on wraparound into real-time control room operations is not a feature of CAISO operations, though the placement of California Energy Resources Scheduling (CERS) personnel, who purchased electricity during the state's electricity crisis and who were employed by the state's Department of Water Resources, demonstrates that such an expansion did take place at one time in CAISO.

In another infrastructure, de Bruijne (2006) provides a fascinating case study of the Dutch mobile telephone network operator, KPN Mobile. An important part of the study was based on KPN control room interviews and observations (see also de Bruijne and van Eeten 2007). He describes a network, which by all accounts should not be as reliable as it became. As in the CAISO case study (to which de Bruijne also contributed), the Dutch telcoms sector has been subject to a great deal of turbulence and volatility due to market restructuring and technological change, among other factors. Yet de Bruijne found that, even in the face of that change,

those who operate these systems have so far been able to cope with these conditions and still provide highly reliable services. Through unconventional methods that at first might appear to have eroding consequences for reliability, KPN Mobile's operators provided highly reliable services. . . . [T]wo responses particularly stand out. First is the increased importance of real-time operations for the maintenance of reliability. Second is the role of control room operators as a key asset for maintaining high reliability of service provision in an institutionally fragmented industry. . . . [C]reative and experienced operators used the markets and options that new technologies provided to cope with the rapidly changing system conditions that sprouted from these developments and thus succeeded in maintaining the real-time reliability of KPN's mobile services (2006, 360).

Moreover, strategies that KPN operators and their support staff use to cope with the system volatility in the Dutch telcoms sector match those discussed in this book (Table P3.1).

Careful observation of other infrastructures also reveals the importance of reliability professionals for high reliability performance. In air traffic control, the maintenance of aircraft separations is a task that must be devolved to the skill-level of the individual controllers and their responsibility for their sectors of airspace (Sanne 2000). In a discussion of air traffic control reliability, a senior official of the FAA remarked, "The ultimate guarantee of safety in this

TABLE P3.1.
Networked reliability conditions in the Dutch
mobile telephone industry at KPN Mobile

Real-time management	Increased complexity and unpredictability of failures forces system operators to maintain system reliability in (near) real time
Heightened sensitivity to additional (real-time) information	A continuous search for additional (real-time) information from outside to raise awareness, detect potential disturbances, and increase responsiveness
Operator experience as fallback during real-time, up-tempo operations	Individual operator experience to maintain reliability in the face of real-time volatility and complexity
Real-time, on-line experimenting	Small-scale testing of new equipment and responses to reliability threats
Redefining reliability	Existing reliability rules and conditions are reexamined and redefined

SOURCE: Adapted with permission from Table 9-1 in *Networked Reliability: Institutional Fragmentation and the Reliability of Service Provision in Critical Infrastructures,* by Mark de Bruijne, Delft University, 2006.

system is the individual controller" (Schulman 1993b, 41). In extended interviews and observations, the importance of the special ability of controllers to see three-dimensionally the air traffic displayed on their two-dimensional radar display screens was consistently highlighted. Those who do not possess this ability are quickly weeded out in training.

Medical treatment facilities constitute an infrastructure that has also received a great deal of attention recently in relation to reliability. A landmark report, *To Err Is Human* by the Institute of Medicine (2000), estimated that up to ninety-eight thousand deaths a year could be occurring from preventable medical errors. The report argued that it was system-level and design failures rather than individual errors that were responsible for most of these deaths.

In fact, a great deal of research in medical treatment facilities also highlights the importance of the individual skills of medical personnel—surgeons, attending physicians, anesthesiologists, and nurses—in preventing errors and enhancing safety in medical treatment (see also Leonard, Frankel, and Simmonds [with Vega] 2004). Some of this research has focused on their skills in diagnostics (Gawande 2003; Groopman 2007); others describe the importance of sensitivity to error among these key personnel (Bosk 2003; Rosenthal and Sutcliffe 2002; Schulman 2004; Klein, Ziegert, Knight, and Xiao 2006).

Other work, while not directly about reliability professionals, addresses the organizational capacities that are necessary for high reliability management. The literature on crisis management is increasingly focusing on the need for resilience, much in the same way as has been developed in this book. What Boin, 't Hart, Stern, and Sundelius (2005, 36–37) term "resilient organizations"

develop an impressive capacity to grasp crisis dynamics. These organizations often work in extremely fast-paced and potentially deadly environments—think of military, police, and rescue service organizations—-but they also exist in high-technology environments (nuclear power and chemical plants). . . . They force themselves to continuously probe their situational assessments—identifying indicators that can be monitored or "tests" devised to provide warning bells to go off should the initial assessment be off the mark. . . . Resilient organizations have created a culture of awareness: all employees consider safety the overriding concern in everything they do. They expect crises to happen. . . . These organizations do not expect employees to rely on their intuition alone (even though leaders of these organizations understand the importance of expert intuition); employees are constantly trained to look for glitches and troubling signs of escalation.

Because most of the critical infrastructures are privately owned, operated, or managed, it is not surprising that private firms and organizations prize just such resilience for many of the same reasons as already discussed. In his *The Resilient Enterprise* (2005, 264–265), Sheffi identifies interrelated cultural elements common in his case examinations of "flexible and resilient" organizations in ways that underscore the importance of operators and managers:

FIRST: Continuous communications provide workers with both a general "state of the company" and with real-time situation reports so actions can be taken quickly and in context.

SECOND: In time-constrained situations, there is deference to expertise (whether or not the expertise comes with a title) . . . and strong teamwork, helping to identify the right response without delay.

THIRD: Distributed power allows employees to take timely action.

FOURTH: Management is very much involved with operations and is knowledgeable and experienced. In fact, it is management's knowledge of the operational environment that makes it confident enough to let [operators] respond with no supervision in cases in which a fast action is called for. It also lets management lead by example.

FIFTH: Hiring and training practices lead to passionate employees who can be entrusted with power to act when called upon by special situations.

SIXTH: The organizations are conditioned to be innovative and flexible in the face of low-probability/high-impact disruptions through frequent and continuous "small" challenges.

It is noteworthy that both Sheffi (2005) and Boin, 't Hart, Stern, and Sundelius (2005; also Boin and McConnell 2007) draw from the literature on high reliability organizations in their findings.

Last but not least, Gary Klein has described the skills of firefighters in terms that closely parallel our description of reliability professionals (Klein 1998). His analysis of the ability of experienced experts to "see the invisible"—to embrace a larger picture beyond the details of a real-time incident—is just what's involved in the pattern recognition skills we have described among control operators. The model he outlines of "recognition-primed" decision making, in which firefighters formulate and refine options as part of their understanding of a decision problem itself, mirrors the scenario-formulation process we have observed in the CAISO control room.

ALTHOUGH THE EXTANT LITERATURE is supportive, we do not pretend it is definitive. In fact, one of the main arguments of this book is that more empirical work needs to be done on how reliability professionals manage our critical infrastructures. Thus our concluding chapters and the implications drawn there are based on the assumption that reliability professionals are core to protecting our large technical systems, not just for electricity but also for water, financial services, and telecommunications (which depend so centrally on electricity[2]) in the ways detailed in Parts One and Two of this book. We feel confident about this assumption, and we look forward to new research that probes and develops it in new ways. Let us now turn to the major implication we draw from our research and review of the literature.

WE ARE VERY WORRIED about the future of society's critical infrastructures.

We are worried about how well high reliability management is understood by those who say reliability must be assured but who fail to learn about the duties and responsibilities of the professionals described here. We have heard and read rubbish from designers, economists, and regulators about what it takes to manage systems about which they and their staff have no operating knowledge. Politicians and economists continue their design attacks on the California electric grid with proposals that would fail every reliability test given them. The next chapter examines the kinds of failed tests that concern us, and the chapter after speaks to where we must look for leadership if society is to take the reliability of its critical infrastructures as seriously as it takes their failures.

To frame the next two chapters and provide the context for the other implications drawn, we summarize for ease of reference four major findings from our case study:

• Reliability issues were not a significant enough part of the debate and policy design process surrounding restructuring in California. The grid was treated essentially as a "black box" in the analysis of players and their market relationships. The full transaction costs of market exchanges were not and still are not factored into pricing and rates because the market players have not internalized the reliability burden imposed by these exchanges. Yet service and grid reliability are thoroughly taken for granted in the market strategies of generators and end-users. The one-time motto of electricity restructuring, "Reliability Through Markets," should actually have been, as it continues to be, "Markets Through Reliability."

- The major professions that informed policy debate and design over restructuring—economics, engineering, and policy analysis—failed to bring reliability issues into proper perspective. Economic analysis treated reliability management as a free good for market participants. Consulting engineers failed to address the organizational and institutional dimensions of providing reliability. Policy analysts failed to appraise policymakers and the public of just how seismic a departure from settled knowledge and institutional practice restructuring was and the potential for displacement of the risks onto the public in this policy experiment. We saw massive social engineering by professions whose members disdain thinking of themselves as social engineers.

- While much attention given to CAISO and its regulators focuses deservedly on improving the electricity system so as to avoid blackouts or, barring that, to ensure restoration of load as quickly and safely as possible, such attention looks only at the right side of the high reliability management framework, namely, just-in-case and just-this-way performance modes, in which system volatility is low. Our research indicates that, notwithstanding improvements (if not precisely because of them), high system volatility and uncertain options variety must be expected in the future and addressed. This means more attention should be given to improving performance on the left side of the framework in terms of enhancing just-in-time performance while avoiding prolonged stays in just-for-now performance.

- Finally, because reliability management has been and continues to be a neglected perspective in electricity system design and redesign, special efforts must be made to incorporate reliability issues into future changes, both in policy and in management. The failure of formal macro-designs ironically may make operational redesigns all the more necessary and feasible.

CHAPTER 12

RELIABILITY AT RISK

T HE QUESTION THAT ANY MAJOR DESIGN
and technology proposal for critical infrastructures must
address is, Does it pass the reliability test of increasing options, decreasing
volatility, or increasing the cross-performance maneuverability of that infrastructure's control operators?

How could any such proposal be assessed in practice? Consider the current
proposal by economists for real-time residential metering to enable consumers
to adjust their energy use in light of their own needs and current market prices
and incentives for energy usage. Such a capability is under active development
in California at the time of writing, through the involvement of many stakeholders, most notably the California Energy Commission (Stockman, Piette,
and ten Hope 2004).

The idea is to motivate energy conservation during peak hours when
prices would be driven up by demand. This, it is hoped, would distribute

power use more evenly during the course of a day and help to flatten out load peaks. Reduced peak demand would also reduce the load on transmission lines and make the job of the control operator easier.

Sounds reasonable . . . until you start parsing the design assumptions underlying residential metering. Proponents ask us to assume that (1) real-time metering, ultimately for all electricity customers, can be successfully implemented; (2) the meter technology will have been thoroughly tested, integrated, and stabilized before being introduced, that is, there are no computer or software glitches (unlike the experience with software used in the restructured network since its inception); (3) what works as a prototype project in Puget Sound or some part of Georgia will work throughout California; (4) there will be no common-mode or cascading failures with meters and automated response control systems; and, in short, (5) everything happens without increasing system volatility.

System volatility still could increase with real-time residential metering, notwithstanding the assumptions. Suppose, for example, that a hot afternoon is forecasted, when it is obvious to large electricity consumers and a large number of households that price spikes in electricity costs can be expected in the afternoon with high power demand for air conditioning. Suppose that quite rationally each consumer decides to run air conditioners early, when power is cheaper, to cool down rooms and houses to unusually low temperatures in order to prepare for the later heat. The effect could be a shifted (rather than reduced) demand peak early instead of late in the day. It might be that this early demand would itself be softened by rapid price rises and subsequent responses in a computerized system as consumers "learn" about the actions of other consumers. But what about the counter-learning strategies and their impacts? After several episodes, many consumers may go back to the early air conditioning strategy, convinced that other consumers have been scared off from trying it because of the earlier price rise. (As someone once said of a popular restaurant: "It's so crowded no one goes there anymore!")

The point here is that a large-scale retail market with real-time price fluctuations could create a great deal of unpredictability or uncontrollability in demand and thus the electrical load facing CAISO control room operators. Isn't this exactly what we see in stock and credit markets with their real-time pricing and millions of participants? Would anyone say these markets have reduced volatility for traders and brokers?

In brief, it is conceivable that real-time residential metering could *reduce* the reliability of load forecasts, add volatility, and challenge the prior pattern recognition of control operators. We could extend our reliability test further and question whether residential metering might actually reduce control room *options* to respond to system volatility. In a new residential metered arrangement, more and different information will be needed by power sellers in order to determine whether providing additional supply is financially worthwhile. In the absence of that information, just as stock sellers might wait to see how a market is behaving (or set a trigger price) on a given day before deciding to offer up their shares, some suppliers might prefer to adopt a wait-and-see strategy before investing in or offering new supplies to CAISO. Resources thus go down at a time when operators may well need them for reliability management. Other questions about residential metering are as easily raised; for instance, just what level of education would be needed for consumers to understand a residential metering system and decide how to formulate their own energy use strategy?[1]

Last but not least, another question needs to be considered. Currently the information to be provided through real-time residential metering will be solely about market price and current demand. Nowhere will information be signaled to consumers about the reliability challenges faced in having to move to just-in-time or even just-for-now conditions, that is, by having unpredictable load curves. There are enormous pricing problems in trying to reflect high reliability management requirements for critical infrastructures in individual consumer decisions.

Do questions such as these mean there should be no real-time residential metering? Of course not. Certainly there is nothing wrong with demand management in principle; the devil is in the specifics (see Barbose, Goldman, and Neenan 2004). If the tools were there for demand management, it might indeed give control room operators more options to be reliable, such as more protection against peaks. It may be that designers and advocates will have answers to the questions raised here as well as to others. It may be that such questions will produce even better designs and technology for coordinating consumer behavior.

The point, though, is that we found no one in the economics, engineering, or policymaking professions during our six years of research and interviews even remotely approaching the kind of analysis we have proposed here. Nor are we aware of any prior consultation about the merits or demerits of residential metering with actual control room managers and operators we have observed over these years. Instead, some economists seem to think that calculation of consumer surplus associated with real-time metering options constitutes the only real "implementation analysis" needed (for example, Borenstein 2004). Such avoidance of or indifference to the hard issues is all the more lamentable, given that some proposals—such as Resource Adequacy—actually do pass the reliability test and end up stabilizing networked operations. Unfortunately, that operational redesign was necessitated by ill-conceived policy and regulatory designs for restructuring. Let us hope there will be a more careful reliability analysis accompanying the enthusiasm for real-time residential metering.

THE STAKES ARE VERY, VERY HIGH in seeing the difference between conventional approaches to design and technology on one side and the very special kind of management required for reliability in critical infrastructures on the other.

Another example of the stakes is homeland security. According to some experts, the best way to reduce the vulnerability to attack of our critical infrastructures—electricity, water, telecommunications—is to redesign them by decentralizing facilities. "The current electricity system with its large, central generators and long transmission and distribution lines is inherently vulnerable," concludes one analysis (Farrell, Lave, and Morgan 2002, 55). "In contrast, a system with many small generators, located at large customers or in neighborhoods, could be made much less vulnerable."

An early report, *Making the Nation Safer* (National Research Council 2002), found that for a variety of infrastructures, including energy, transportation, information technology, and health care, "interconnectedness within and across systems . . . means that [the] infrastructures are vulnerable to local disruptions, which could lead to widespread or catastrophic failures." At issue for *Making the Nation Safer* and subsequent analyses is the notion that critical infrastructures have discrete points of vulnerability or "choke points," which if subject to attack or failure threaten the performance of the entire infrastructure. Key transmission lines upon which an electrical grid depends, or single security screening portals for airline passengers, or a single financial audit on which countless investment decisions are predicated are points of vulnerability that can lead to cascading failures throughout an entire infrastructure. The more central the choke point to the operation of a complex system, so this logic goes, the more dependent and susceptible the system is to sabotage or failure.

From the designers' viewpoint, the only remedy to the vulnerability of complex, tightly interconnected systems is to redesign them to simplify or decouple the parts and render them less interdependent. Such a strategy would decentralize power grids and generators, making smaller, more self-contained transmission and distribution systems. In business continuity programs of large companies, the decentralization strategy has involved diversifying the hardware systems and geographically dispersing the databases and data processing.

Although the current view sees choke points as the problem and decentralization or going off the grid altogether as the answer, the issue is not at all clear-cut from the perspective of the reliability professionals running the control rooms for the electrical grid. As we have seen, tight coupling and complex interactions can be system resources as well as sources of vulnerability. Choke points are the places where terrorists will likely direct their attention, but they are also, from an operational standpoint, the places to which control room operators, with their trained competencies in anticipation and resilience, are also most sensitive and attentive.

Indeed, our research suggests that the prevailing view could be stood on its head. Complex, tightly coupled systems confer reliability advantages to those control room operators who seek to restore them against terrorist assault. As we have seen, their complexity allows for multiple (equifinal) strategies of resistance, resilience, and recovery after failure. Their tightly coupled choke points enable these professionals to see the same portion of the system as the terrorists, and position them to see it whole against a backdrop of alternatives, options, improvisations, and counter-strategy.

During the electricity crisis, for instance, California's grid was tightly coupled around the major choke point of Path 15, the high voltage line connecting the northern and southern grids. Path 15 remains a target for terrorists who might see it as a single point for great disruption. But the professionals who manage the grid on a real-time basis have already experienced a variety of hardships with Path 15—overheating, which limits its capacity to carry power, and congestion, which blocks feeder lines into and out, as well as failures along sections of it caused by fires, storms, earthquakes, and more. Managers and operators have fashioned multiple solutions to these problems—ranging from rerouting power along alternate paths to calling on backup generation in localized areas to strategic load reductions when necessary. They can counter threats to Path 15's vulnerability by calling on this experience, their overview of the entire system, and their real-time armory of options and improvisations. Indeed, it is difficult to imagine a case in which real-time improvisation and adaptation with respect to known choke points would not be a major part of any control room's counterstrategy to a determined terrorist attack.

In contrast, those decoupled, or loosely coupled, decentralized systems present many more independent targets. Terrorists could strike anywhere and, while they may not bring down major portions of the grid, they would still score their points in the psychological game of terror and vulnerability. In addition, the local managers would probably not have a clear picture of what was happening overall, nor would they have as wide a range of alternatives and recovery options available.

These trade-offs between the current centralized system and the alternative redesigns for more decentralized systems are neglected, if not ignored outright, because the designers frequently have no clue whatever that the trade-offs exist. Instead, the nation's first line of defense, the skill base of reliability professionals, is ignored.

In reality, reliability professionals are our first responders in case defenses fail. They are the first ones who should be asked about proposed "improvements" to critical infrastructures. They should be asked how these changes might increase their options to respond to threats or enhance their ability to match options to vulnerabilities. Some answers will doubtlessly be both surprising and enlightening to many designers and leaders of organizations—and they may save lives and critical assets in the process.

SAVING LIVES IN THE FACE OF ATTACK brings us more directly to current approaches to counterterrorism strategy. Given our analysis of high reliability management, much concerns us, but an example will suffice. A recent controversy over whether a tip that terrorists planned to attack subway trains in New York with bombs in suitcases and baby carriages was a hoax or not raises an important question concerning counterterrorism strategy. The question goes beyond whether this particular threat was a hoax to how we should treat tips and terror threats.

A *Wall Street Journal* article ("A Hoax or Not," October 13, 2005) noted a growing number of such false tips, with law enforcement officials obligated to investigate each one. One New York City official, in defending the city's response to the subway tip, contended, "I don't know of any valid information to indicate that it was a hoax." Yet this approach, placing the burden of formal proof on the side of discounting a hoax, is, we believe, far less precautionary than it seems.

The perspective at work among New York officials is quite similar to the precautionary principle argued for by environmentalists. One must try to guard against harm to the environment by assuming the worst-case environmental outcome of human activities. Unless this worst case can be formally and definitively refuted, assume that it is the actual case and shape policy accordingly (also see Sunstein 2007). This orientation to precaution would seem to be a necessary reliability strategy to guard against events and outcomes that we simply do not ever want to happen. Indeed, we have encountered many people in our research who operate to a precautionary standard in their effort to avoid operating in the precursor zone or avoid remaining in just-for-now performance mode for any extended period of time.

But there is an important difference between extreme and formal precaution with respect to a single event and high reliability management over many events. It is impossible to prove that something will not ever happen. More important, to demand formal proof in order to discount each and every worst case—for example, a formal validation for each "hoax"—actually negates the judgment and experience of those field operatives and professionals—be they law enforcement officers or control operators—upon whom ultimate safety and reliability depends.

It is all well and good to say that critical infrastructures should plan for the "worst case scenario." But scenario formulation is only one side of the reliability professionals' domain. You cannot ask them to ignore the other side, their pattern recognition and the anticipations based on those patterns. To be reliable, professionals start with the frequency of occurrence of a hazard as well as the magnitude of such hazards. It is their ability to probe the scenarios, worst and otherwise, *and* the patterns, present or absent, that enables them to promote reliability—if simply to determine if what others take to be the "worst case scenario" is worse enough.

The professionals we have observed would not squander scarce resources of attention or public credibility on threats that in their judgment do not reach a plausible level of concern. They would not attempt to apply some formal standard of proof before they could assess each threat, discount it, and move on. Nor should they be forced to work to such illusory standards. The key to high reliability management is not to try to eliminate all risk in every case; rather it is to manage risks (note the plural) over many potential cases and different performance modes. A formal doctrine that requires validating the falsity of every tip does not constitute good risk management—it simply increases the risks of bad management.

What is missed in the "safety" criterion, and dangerously so, is its direct move from formal macro-principle to an individual micro-case. In doing so, it bypasses the domain of the reliability professionals within which they appraise each individual case in terms of a pattern of cases and scenarios. To compel the professional to prove that each and every tip is not a hoax robs the reliability professionals of the ability to use their skills to manage risk. In case it needs saying, making public one's macro-design principle also encourages counterstrategies (such as planting multiple hoaxes to diffuse attention), which in turn can further undermine the ability of the reliability professionals to manage risk. We could not agree more with Steven Kelman, professor of public management and government administrator, who argues when writing about the 9/11 Commission: "'Good' performance isn't enough. Counterterrorism organizations need to be high-reliability organizations of a special kind" (Kelman 2006, 141).

CHAPTER 13

THE FUTURE OF
HIGH RELIABILITY
MANAGEMENT

O N MARCH 23, 2005, BRITISH PETROLEUM'S
Texas City Refinery exploded, killing fifteen, injuring
over a hundred others, and resulting in significant economic impacts. The explosion was one of the worst industrial accidents in recent memory in the
United States and has become the subject of several investigations (McNulty
2006). The most prominent inquiry was chaired by former secretary of state
James A. Baker III, and it issued the "Report of the BP U.S. Refineries Independent Safety Review Panel" in January 2007 (BP U.S. Refineries Independent Safety Review Panel 2007). The Panel report found major problems in
BP's culture of safety, especially in the area of what it called process safety:

Process safety incidents can have catastrophic effects and can result in multiple injuries
and fatalities, as well as substantial economic, property, and environmental damage.
Process safety refinery incidents can affect workers inside the refinery and members of

the public who reside nearby. Process safety in a refinery involves the prevention of leaks, spills, equipment malfunctions, over-pressures, excessive temperatures, corrosion, metal fatigue, and other similar conditions. Process safety programs focus on the design and engineering of facilities, hazard assessments, management of change, inspection, testing, and maintenance of equipment, effective alarms, effective process control, procedures, training of personnel, and human factors. The Texas City tragedy in March 2005 was a process safety accident (2007, x).

What makes this accident of particular note is BP's goal of ensuring a "high reliability strategy" to prevent such accidents:

BP Refining's strategic performance unit aspires to be a "High Reliability Organization" (HRO). An HRO is one that produces product relatively error-free over a long period of time. This initiative began with a meeting of a handful of refining leaders in the fall of 2000. . . . Refining management views HRO as a "way of life" and that it is a time-consuming journey to become a high reliability organization. BP Refining assesses its refineries against five HRO principles: preoccupation with failure, reluctance to simplify, sensitivity to operations, commitment to resilience, and deference to expertise. BP's goal is to inculcate HRO through the refining process and thereby create and maintain a culture in which reliability and safety are core values (2007, 40–41).

Because the Panel report found the Texas City refinery accident to be a failure in the culture of safety, the immediate question is, Just how realistic is an HRO or high reliability strategy for our critical infrastructures generally?[1]

The stakes are exceptionally high in answering this question correctly. It is increasingly common today to recommend a "high reliability" approach to improving organizational resilience in the face of safety and security mandates (for example, Sheffi 2005, Chapter 15). "Reliability is part of brand protection," states the headline to an article on crisis management and business continuity in the face of catastrophe (Tieman 2006, 2). The recommended reliability approaches, however, are frequently grounded in strategies of technological design or expanded markets. Moreover, as just seen in the preceding chapter, decoupled or decentralized systems are assumed to be an answer to providing increased safety and reliability (National Research Council 2002), given that accidents and errors are inevitable in those technologies whose designs make them complex with tightly coupled components (for instance, Perrow 1999 [1984]). But in our view, high reliability management must be grounded foremost in institutions, not technologies or markets.

In his classic *Leadership in Administration,* Philip Selznick argued for the role of leaders in transforming organizations into institutions. The transformation entails establishing ties of social acceptance and dependency upon the functions of the organization (Selznick 1957). The permanency of organizations rests upon this institutional transformation, and organizational leadership is a primary means by which to bring it about.

Organizations that must manage critical infrastructures reliably need to be institutions. They need to enjoy public acceptance, because to perform effectively reliability must be embedded in norms at the heart of their culture and the culture of the society around them. This allows some organizational output to be legitimately foregone on behalf of safety—a nuclear plant can go offline or an air traffic sector can be closed rather than operate under unsafe conditions.[2] As institutions, these organizations are allowed to gain control over important aspects of their task environment, stabilizing a variety of organizational resources such as personnel, materials, and some measure of environmental control.

In contrast, the California Independent System Operator has been confronted with continual shifts in regulatory approaches and standards, and jurisdictional battles between the California Governor's Office and regulators, including FERC and the CPUC. CAISO has faced skepticism, from both the major utilities and independent generators, and resistance to its market management and its grid management directives. It has confronted ongoing challenges to its practices from outside economists, engineers, lawyers, and politicians. It has faced active resistance and illegal market manipulations by private energy sellers, whose proprietary interests have been placed above the larger societal value of grid reliability. CAISO confronts constant changes in market conditions, market strategy, new technology introduced on the grid by the utilities and generators, and even a challenge to the employment status of its control operators by a California court.[3]

Institutions have been described as "highly cherished policy delivery machines" (Boin 2004). As we have seen, for CAISO the biggest challenge to its institutionalization as an organization has been the market and network relationships underlying the restructured electricity it is supposed to deliver. During the six years of our case study, CAISO has been continually attempting to stabilize its relationships with other major players. It has sought new regulatory support to make market participants act in ways to safeguard grid reliability, from RMR contracts to Resource Adequacy requirements. It is undertaking its own market reform program, MRTU, to allow the California electricity markets to work more efficiently with more power transactions in day-ahead instead of hour-ahead and real-time markets. Finally, CAISO has evolved a new business plan with a rate rollback and the promise of faster posting of settlement statements for power schedulers.

Moreover, some efforts have worked. From CAISO's perspective, Resource Adequacy has stabilized the state's deregulated network of generators, transmission managers, and distribution utilities. As we have seen, control room operators moving to the edge of their performance may look the same during two periods of time, but they vary considerably in terms of whether the time period was before or after Resource Adequacy was in place. The drive to make incomplete networks of diverse interests more stable in order to better ensure reliable critical services is not a new one or limited to CAISO.[4]

Notwithstanding these positive improvements, the deregulated grid remains incompletely designed. CAISO still faces environmental resistance and turbulence. Distribution utilities and individual generators still at times resist compliance with CAISO directives.[5] Its relationship with its regulators continues to have tensions.[6] Some FERC officials would have liked to see the end of CAISO altogether and its replacement by a regional transmission organization (RTO). Added to this is the general regulatory uncertainty created by the Energy Policy Act of 2005. The act calls for the creation of national reliability standards across all control areas and the certification of electricity reliability organizations (EROs) to impose these standards. At this writing no one yet knows what the new standards will fully be.

ALTHOUGH WE HAVE IDENTIFIED the threats of formal and technical design to high reliability management, we have also seen how operational redesign can improve network relations for better performance. Indeed, when policy design fails, all we are left with are reliability professionals and operational redesign to ensure that high reliability management continues. Our analysis suggests three specific design principles that further rather than subvert reliability operations.

First, as a matter of principle, every design proposal should pass that reliability test keyed to the work of these professionals. Would the proposal, if implemented, reduce (or least not increase) the volatility faced by operators, who are really frontline defenders? Would it increase (or at least preserve) their options to respond to volatility? Second, as a matter of principle, any design that compels operators to work in a task environment outside their domain of competence in pattern recognition and scenario formulation for prolonged periods of time cannot be expected to produce a reliable system. Third, as a matter of principle, cognitive neutrality should be followed, in which new variables are introduced into high reliability management only if they pose no net increase in the cognitive requirements of their control room managers. The cumulative effect of these three principles is to insist that conscientious designers are those who revise proposals to make them more practicable before proceeding further. The principles compel system designers to learn about the cognitive workload of operators and the system contingencies that cannot be planned for but which must be managed by these reliability professionals case by case.

In this way, CAISO's role as an institution means that design, planning, and regulation initiatives outside the control room must be informed by and be consistent with the unique knowledge base of reliability professionals inside the control room and their immediate support units. Return again to the reliability test. Some control rooms, in nuclear power plants, airplane cockpits, and air traffic control, already have veto power over the introduction of new procedures and software when operators believe these changes threaten to degrade their reliability performance. This veto power should be extended further.

Another road to be taken is to make designers part of the wraparound of support units to the professionals in the middle. That support staff already includes engineers who test their design proposals with operators, as we demonstrated in our case material. Ecologists too are moving into the control rooms (and their wraparounds) of some of our major water supplies in order to better meet their environmental reliability mandates (Van Eeten and Roe 2002).

Why not the same for legislative and regulatory staff through cross training, attendance at training workshops for operators, immersion stays in operations, or longer visiting arrangements with the critical infrastructures they are meant to regulate? The fact that state and federal laws and regulations are being written by legislative and regulatory staff, aided and abetted by fellow-traveling academics, all of whom know nothing of what it takes to manage for reliability in real time, should be cause for widespread alarm.

In general, designers, planners, and regulators must be much more open to eventualities unpredicted by themselves or their staffs and consultants. Lawyers are fond of writing for all known conditions (Kay 2006), and that alone is necessity enough for having reliability professionals to manage the ensuing risks. So too for regulators. They frequently seek to control for the full range of conditions, and it is precisely that mandate for comprehensive coverage that creates the need for a unique domain of competence of the reliability professionals. No set of regulations or laws could ever incorporate or reflect the knowledge required to ensure reliability under all the contingencies that matter precisely for the mandate(s) set in those regulations or laws.

ONE OR MORE of the following strategies, it seems to us, could help rebalance the policy formulation process to include the neglected managerial perspective discussed in this book.

1. When a managerial perspective is key, the best way to obtain it is to ask managers.

Why not ask them to testify at committee hearings regarding proposed legislation that they will be charged to implement? It is crucial for reliability professionals to offer their own assessments of the likely impact of specific design and technology proposals on their ability to shift performance modes when necessary.

2. We should think hard about how to upgrade the professional standing and social role of reliability managers.

Whereas economists, lawyers, and engineers, major players in modern policy design, have clear professional identities, the reliability professionals we describe do not.

"When the public hear the word *dispatcher*," said one shift supervisor, "they think that's what gets you the Yellow Cab. They think anybody out there can do that job." Speaking of the California grid, a CAISO engineer said, "It's become a lot more complicated, and people outside don't understand that. People don't appreciate that you can't stop the grid to fix it."

From our research it is not clear that reliability professionals themselves fully appreciate the special skills and cognitive orientation they possess. It is not easy to "manufacture" or "certify" a professional identity, but it has happened for a variety of occupations (Wilensky 1964). It might well be useful for the organizations in which they work to facilitate cross-communication, if not cross-training among similar personnel in different organizations. It would be of great benefit, we believe, for CAISO controllers and shift supervisors to meet control room operators and supervisors in generating facilities and distribution centers among large customer-based electricity companies elsewhere. One venue for this is the American Power Dispatch Association. Other associations of reliability professionals should be encouraged, perhaps in this country with the support of the American Society of Professional Engineers (ASPE) or the American Society for Public Administration (ASPA).[7]

In the meantime, it must be recognized that long and diverse work experience is core to the training of reliability professionals. Care must be taken not to destroy the potential career paths of these individuals. In California, the bankruptcy of a large utility (PG&E) and its diminished role in controlling its own electrical grid, as well as the substantial downsizing of the workforce among private generators, mean that fewer opportunities exist for people to establish careers in electric power, to work their way through a variety of positions on all sides of the generation, transmission, and distribution system.

In 2001, a senior CAISO official concluded his interview with us by saying, "I am worried about a skilled operator interface. Like how do we train the dispatchers in the control room? We are short on that. Look at the control room. [Three people there] are all retirees. I need the people with the knowledge, skills, and completeness to handle the job in the future. . . . Now we are using retirees!" In 2006, retirees are no longer the majority, and the recent CAISO reorganization has reduced CAISO staff overall, including those in the control room.[8] What the control room needs are people with utility experience, but it also needs people who know what it means to operate at the different scales and through different critical phases created by restructuring and its in-built volatility.

3. We need better pricing of high reliability.

"No one is pricing for reliability changes," a senior grid manager at CAISO said in 2004, by which he meant the continuing difficulties over determining just what costs CAISO and the distribution utilities bore individually and shared in operating the grid. It might be said of reliability professionals, "They get paid to do what they do, so what's the problem?" The problem is that price fluctuations in electricity simply do not reflect the reliability edges we described in Chapter 11. Movements in prices do not match movements to the edge, and thus do not reflect the value-added that operators provide by having to compensate for the gaps between design and the actual challenges of moment-to-moment operation. Briefly, current prices are a poor reflection of how close on occasion reliability professionals are pushed to the edge of their reliability envelope. We even wonder if CAISO job descriptions for control room operators could capture these important operational distinctions.

In one sense, the neglect of clear reliability pricing occurs because the high reliability of a critical infrastructure amounts to a public good different from the market goods and services that people purchase (see Auerswald, Branscomb, LaPorte, and Michel-Kerjan 2006). This has been more than amply demonstrated in the California electricity crisis by the willingness of the state to spend its billions in public funds to maintain high reliability.

But there is an additional sense in which prices do not attend to the requirements of high reliability management. As we have seen, maintaining always-on reliability requires the ability of reliability professionals to respond to a wide variety of shifting conditions and to bring their critical skills across diverse performance modes. This is what we have described as high process variance required to maintain low output variance, in other words, the balance of load and generation. Yet these performance modes have varying ranges of difficulty. In addition, certain grid conditions and market transactions push reliability professionals into a precursor zone closer to the edge of their skills and competencies.

The efforts required to stay out of the precursor zone are not reflected in any variable costs currently applied to market transactions regarding electricity. For example, some transactions (scheduled in the hour-ahead or which impose high congestion management requirements) could well be priced differently in relation to the heavier burden they impose on operators in order to stay out of the precursor zone. Because these burdens are not fully priced, high reliability management is undervalued in market transactions. When undervalued, reliability management falls victim to societal underinvestment.

An important first step in the full-cost pricing for reliability entails the development of indicators of movements toward edges in the balance between operator skills and task requirements. We undertook just such an indicators project at CAISO, described in Chapter 11. Quantitative indicators provide evidence for connecting distinct inputs or challenges to actual shifts in reliability performance variables. In this way, they provide evidence for pricing services or market behaviors that push operators into precursor zones with respect to their performance edges.

Full-cost pricing of reliability would not mean that high reliability has itself become a marketable or fungible item. Rather, what would be better priced are the costs and some important risks to high reliability management. These would now be priced in a way that takes into account the empirical understanding of changing conditions and their impacts on the way in which reliable grid management operates.

Eventually, where costs are better understood, more careful judgments regarding reliability become possible. All participants in a complex system such as a critical infrastructure can be made to understand how specific conditions and their own behavior can have measurable effects on important dimensions of reliable performance. This in turn allows more accurate pricing of reliability as a significant component of interorganizational transactions and market exchanges.

4. Ultimately, the question of the balance between design and management in the reliability of critical infrastructures will rest, we believe, on the evolution of professional ethics.[9]

The promulgation of policy designs that ignore institutional challenges and requirements is not simply narrow and short-sighted; it is irresponsible. Our increased dependency on the reliability of critical infrastructures as well as the other services dependent on them heightens the consequences of design errors in the policies applied to the infrastructures. Currently, professional designers suffer few consequences for their design errors. In fact, the consequences of their errors are displaced onto clients, customers, and the general public as reliability becomes more uncertain and the rectification for policy error more costly.

At a minimum, professional standards should be a constraint upon willfully negligent public prescriptions regarding policy. It is a professional responsibility to exercise a prudent forbearance about the potential for errors, both of omission and specification, in formal designs urged upon our public resources. Policy analysts and professional designers should be, as policy analyst James Anderson once contended, "partisans of the neglected perspective."

Fortunately, some macro-level designers are aware of the problem and doing something about it. Two of the "Hannover Principles" for design sustainability, prepared for the 2000 World Trade Fair in Hannover, Germany, read as follows:

4. Accept responsibility for the consequences of design decisions upon human well-being. . . .

8. Understand the limitations of design. No human creation lasts forever and design does not solve all problems. Those who create and plan should practice humility . . . (Thackara 2006, 25).

We can only hope to see a like movement take place throughout economics, engineering, and policy analysis.

THESE CONSIDERATIONS have larger implications for the challenge to public institutions inherent in modern policymaking. An important conclusion of our analysis is that policymaking should be closely linked to the management of institutions in order to be successful. After all, for policy to be successful it must be managed and managed reliably. This means that policy designs must address managerial requirements if they are to reap the benefits of reliable management. It also means that policy, particularly public policy, should seek to avoid those problem domains where the skills of reliability professions cannot be brought to bear. Of all the lessons for policymakers highlighted throughout this book, perhaps the most important one is forbearance (see also Roe 1994).

This issue clearly deserves thoughtful and extensive analysis beyond the scope of this book. But to put it briefly and plainly: because high reliability management is important to the effectiveness of a wide variety of policy undertakings, it follows that policymakers should strive to stay out of problem domains (issue areas), where it is not possible to define errors and thus identify and manage risks, and where uncertainty is so great there is little or no chance even to plan the next step ahead. These domains are toxic to high reliability management.

One thinks immediately of the "fog of war." It is here where pattern recognition and scenarios are their thinnest, as unpredictability and uncontrollability—or what Clausewitz called "friction"—are at their highest. "This tremendous friction," Clausewitz wrote in explicating the fog of war, "is everywhere in contact with chance, and brings about effects that cannot be measured" (Paret 1986, 202–203).

So too when the wars are declared in such broadly defined problem domains as "poverty" or cancer. Here the lack of clear or convincing performance standards and cause-effect models is endemic. Sargent Shriver, who took the lead in the War on Poverty, realized early on that he and his fellow policymakers were in over their heads when it came to eliminating poverty (Marris and Rein 1969; Schulman 1980). Since 1972, after the War on Cancer was declared, fewer than 0.5 percent of the study proposals to the National Cancer Institute have focused primarily on metastasis; yet it is the process of metastasis, not localized tumors, which kills cancer victims 90 percent of the time (Leaf 2004).

When problem domains are defined very broadly, policymakers end up in the purgatory of having no theory with which to connect their case, no model with which to establish that this cause leads to that effect, and that effect will have these impacts. To define problem domains so vastly, in our terms, is to work in a space literally undefined by knowledge and unbounded to a scale relevant to reliability. It is to choose to operate where even macro-designs and micro-operations, let alone pattern recognition and scenario formulation, provide little guidance. Policymakers in these problem domains are understandably hounded by the thought that the worst case scenario is one beyond their wildest dreams. But from the perspective of reliability professionals, operating in these problem domains *is* the worst case scenario.

It could be argued that not all policy domains need have as their center of gravity high reliability. Yet who can justify entering (or remaining) in a policy domain in which reliability road marks make no sense or have no significant role? The focus on reliability brings discipline into management. It is the reality check of having to recognize systemwide patterns and formulate case-specific scenarios in ways that can translate into reliable critical services. Reliability compels managers to identify what is or is not error, to balance tolerance and intolerance to the errors identified, and to be resilient in the face of errors while at the same time planning the next step ahead. And even then, success is never retrospectively guaranteed for the future ahead.

It is not as if policymakers have to intrude into domains where management is not welcomed. In many cases, policymakers could and should choose to concentrate policy in domains where reliability professionals can and are operating. In these domains, as we have seen, reliability professionals can send policymakers precursor signals indicating when moves are to the edge of reliability. For example, reliably maintaining military bases is far different from expecting to create democracy where it has never existed so as to secure economic growth and political stability through unspecified best practices for doing so. Reducing poverty for the aged is far different from vanquishing poverty; reducing the incidence of non-Hodgkin's lymphoma is reliably achievable in ways that eliminating other cancers is not. In each instance, policymakers and policy managers have an error horizon against which errors can be identified and balanced.

In domains where reliability professionals can and do operate, policymakers can be assured that management is taking place within bandwidths and can avoid any lock-in of a single theory or pattern to a single scenario with no room to adjust or maneuver in light of unpredictable or uncontrollable events and conditions. Indeed, as we have seen, policymakers can formulate policy knowing that design errors are to be buffered by reliability professionals. For policymakers instead to intrude into problem domains where ignorance and uncertainty are maximized, while at the same time ignoring domains where reliability professionals can anticipate the next step ahead, is inevitably a dangerous and unthinking undertaking.

To put it too briefly, policymakers must be sensitive to those domains where policy does not work. It is here that good judgment leaves progress to history and evolution rather than policy and management. Reliability professionals can help in this discrimination, but they need to be engaged to help identify no-go areas for policymakers. Doing this together is necessarily the best way, we believe, because it is the only way to reliably foreclose problem domains recalcitrant to policy and its management.

WE HAVE ARGUED that we need to learn to distinguish problem domains in which public policy does not work. At the same time, we are currently in the process of losing an understanding of where public policy will appropriately work. In *Decline of the Public*, David Marquand argues that there has been a loss of important distinctions between the public and private "domains":

The public domain is, in a special sense, the domain of trust. . . . The goods of the public domain must not be treated as commodities or surrogate commodities. Performance indicators designed to mimic the indicators of the market domain are therefore out of place in the public domain and do more harm than good. By the same token the language of buyer and seller, producer and consumer, does not belong in the public domain. People are consumers only in the market domain; in the public domain they are citizens. . . . Professions, professionalism and the professional ethic are inextricably linked to the public domain. . . . To carry out their duties, professionals must have the autonomy to exercise their judgment as they see fit (Marquand 2004, 135–136).

Marquand's argument about the decline of the public domain can also be an argument about the decline in regard among policy designers for the weight of public institutions and their management. Unless we find ways to cultivate a managerial perspective in the crafting of public policy, the world of policy design will continue to be an ever-more hostile environment for the successful and reliable management of public institutions.

We leave you with two alternatives for the management of critical infrastructures. One lays out the implications of business as usual. We foresee increasingly close edges and failures of performance in the large technical systems that are ever-more interdependent, with ever-enlarging consequences of failure. We foresee misdirected investments in technological fixes and formal planning at the neglect of investment in resilience and recovery. We see ever-more elaborated markets and technology that, through their neglect of high reliability management, cause the very errors they seek to correct.

Any successful alternative scenario must, we contend, incorporate the experience base of high reliability managers. It is, as we have said, one stupendous irony that, while all manner of economists and market evangelicals call for greater efficiencies in critical service provision, the most underutilized resource we have as a society are the reliability professionals who filter out big mistakes before they happen. We must tap that resource soon. Only by doing so will we have an alternative scenario to that posed in the news report at the outset of this book.

REFERENCE MATTER

RESEARCH METHODS

W̅E̅ RELIED ON MULTIPLE METHODS, DOCUMENTS, and key informants to identify and cross-check our findings. A two-phased study for this research was adopted. In early 1999, we and our colleagues reviewed the literature on deregulation of the energy sector, with special reference to California's electricity restructuring. During this initial phase, we identified the California Independent System Operator (CAISO) as the focal organization for primary research. We approached key officials there and received permission to interview staff in and around the CAISO main control room. Key informant interviews proceeded by the snowballing technique, in which new interviewees were identified as "people we should talk to" by previous interviewees, until we reached a point at which new interviewees were mentioning the same or similar problems and issues heard in previous interviews. An interview questionnaire was used throughout (allowing open-ended responses and follow-up), with face-to-face interviews typically lasting an hour or more. A major portion of the research is from direct observation, watching what control room operators and engineers did in their jobs through many shifts in performance conditions over the years. This included sitting at their console stations, identifying their various monitors and displays, and querying their operations as they undertook them.

The first phase of our research was undertaken with our colleagues at Delft University of Technology, particularly Michel van Eeten and Mark de Bruijne. We undertook the bulk of investigations between April and December 2001. Sixty interviewees were identified and interviewed: thirty-three in and around the main control room of the ISO; eight at PG&E (in and around their Transmission Operations Center and the Operations Engineering Units); five with a large generation supplier (a senior generation official and control room operators in one of its large California plants) and a private market energy trading dotcom; and fourteen others located in the California Governor's Office, California Public Utilities Commission, California Energy Commission, Electric Power Research Institute, Lawrence Berkeley National Laboratory, University of California, Berkeley, and Stanford University. We were also able to observe control room behavior at a high point of the electricity crisis in April–May 2001.

Our interviews and observations focused on control rooms. Earlier HRO research as well as later investigations (Van Eeten and Roe 2002) found control rooms to be the one place where HRO features were visible across a wide range of behavior, namely technical competence, complex activities, high performance at peak levels, search for improvements, teamwork, pressures for safety, multiple (redundant) sources of information and cross-checks, and a culture of reliability, all working through tightly coupled, sophisticated technologies and systems.

Our research was written up as a report on the California electricity crisis and restructuring, and was published in late 2002 by the Electric Power Research Institute (EPRI) on behalf of the California Energy Commission (Roe, van Eeten, Schulman, and de Bruijne 2002). As part of the follow-up, we returned to CAISO periodically from 2002 onward. In 2004, two of the original team—the authors of this book—were requested by the chief operating officer of CAISO to renew more intensive research. He was concerned about explaining a control room incident that had happened in March 2004. We in turn proposed to look more closely at the cognitive skills of the control room operators, and their reliability envelope, during this second phase, which ran from July 2004 to the time of writing (mid-2007). Overall, we talked to at length or formally interviewed at least one hundred CAISO staff and others and spent an equivalent number of hours in direct control room observations. CAISO assigned us an intern for our indicators research during summer 2006, who compiled the database and undertook the initial run of statistics, using the data-analysis package in Excel. These data provide the basis of our findings in Chapters 6 and 11.

DESCRIPTION OF PRINCIPAL FEATURES OF HIGH RELIABILITY ORGANIZATIONS

(See Figure 4.1)[1]

HIGH TECHNICAL COMPETENCE

High reliability organizations (HROs) are characterized by the management of technologies that are increasingly complex and which require specialized knowledge and management skills in order to safely meet the organization's peak-load production requirements (Rochlin 1993, 14; LaPorte 1993, 1). What this means in practice is that the organizations are continuously training their personnel, with constant attention devoted to recruitment, training, and performance incentives for realizing the high technical competence required (Roberts 1988, Figure 3; LaPorte 1996, 63). To do so means not only that there must be an extensive database in the organization on the technical processes and state of the system being managed, but that this "database" includes experience with differing operating scales and different phases of operation—the proposition being that the more experience with various operating scales and the more experience with starting and stopping risky phases of those operations, the greater the chance that the organization can act in a reliable fashion, other things being equal (LaPorte 1993, 3; Perrow 1994, 218).

HIGH PERFORMANCE AND CLOSE OVERSIGHT

Technical competence in an HRO must be matched by continual high performance. The potential public consequences of operational error are so great that the organization's continued survival, let alone success, depends on reliably maintaining high performance levels through constant, often formal oversight by external bodies. As Rochlin (1993, 14) puts it, "The public consequences of technical error in operations have the potential for sufficient harm such that continued success (and possibly even continued organizational survival) depends on maintaining a high level of performance reliability and safety through intervention and management (i.e., it cannot be made to inhere in the technology)." Accordingly: "Public perception of these consequences imposes on the organizations a degree of formal or informal oversight that might well be characterized as intrusive, if not actually comprehensive." LaPorte (1993, 7) adds, "Aggressive and knowledgeable formal and informal watchers [are] IMPORTANT. Without which the rest [in other words, high reliability] is difficult to achieve."

That "oversight" does not mean strict supervision of personnel within the HRO. In fact, overly close supervision is inimical to achieving high reliability (Schulman 1993a). Rather, the oversight in question typically comes from external bodies (such as regulators) that demand high reliability from the HRO's senior managers, who in response allocate resources to achieve that reliability.

CONSTANT SEARCH FOR IMPROVEMENT

A feature related to high technical competence and constant monitoring is the continued drive to better HRO operations. Personnel constantly strive to improve their operations and reduce or otherwise avoid the hazards they face, even when—or precisely because—they are already performing at very high levels. "While [HROs] perform at very high levels, their personnel are never content, but search continually to improve their operations" (Rochlin 1993, 14). They seek improvement "continually via systematic gleaning of feedback"

(LaPorte 1996, 64). Notably, the quest is not just to do things better, but to reduce the intrinsic hazards arising from the activities of many HROs (T. R. LaPorte, personal communication, 2000).

HAZARD-DRIVEN ADAPTATION TO ENSURE SAFETY

HROs face hazards that drive them to seek adaptive flexibility as a way of ensuring safety. "The activity or service [of these HROs] contains inherent technological hazards in case of error or failure that are manifold, varied, highly consequential, and relatively time-urgent, requiring constant, flexible, technology-intrusive management to provide an acceptable level of safety [that is, reliability] to operators, other personnel, and/or the public" (Rochlin 1993, 15). The more hazardous the operations, the more the pressure to ensure the high reliability of those operations. Weick and Sutcliffe (2001, 14) characterize this flexible adaptation as resilience: "The signature of HROs is not that it is error-free, but that errors don't disable it."

OFTEN HIGHLY COMPLEX ACTIVITIES

Not unexpectedly, the more complex the actual operations and activities performed, that is, the more inherently numerous, differentiated, and interdependent they are, the greater the pressure to operate in a highly reliable fashion (for example, Rochlin 1993, 15). What this means in practice is that HROs often find it "impossible to separate physical-technical, social-organizational, and social-external aspects; the technology, the organization, and the social setting are woven together inseparably" (Rochlin 1993, 16). In such an organization, its technology, social setting, and units are extremely difficult to tease apart conceptually and practically. Such complexity characterizes many activities of many HROs. Note the qualification "many." Not all activities in an HRO are complex (T. R. LaPorte, personal communication, 2000). The point here is that the more complex (and the more hazardous) the operations of an organization in combination with the other features discussed earlier and further on, the greater the pressure to manage in a highly reliable fashion.

HIGH PRESSURES, INCENTIVES, AND SHARED
EXPECTATIONS FOR RELIABILITY

HRO activities and operations must meet social and political demands for high performance, with safety requirements met in the process and clear penalties if not (Rochlin 1993, 15). One way to do so is to ensure that those who provide the services work and live close to the system they manage—they fly on the airplanes they build or guide, they live downwind of the chemical plants they run or on the floodplains they manage, and their homes depend on the electricity and water they generate (Perrow 1994, 218).

CULTURE OF RELIABILITY

Because the HRO must maintain high levels of operational reliability, and safely so, if it is to be permitted to continue to carry out its operations and service provision, a culture of reliability comes to characterize the organizations (Rochlin 1993, 16; Roberts 1988, Figure 3). This means in practice that the organizations often exhibit clear discipline dedicated to ensuring failure-free, failure-avoiding performance (LaPorte 1993, 7). Such a culture does not mean the organization is sclerotic with respect to formal safety regulations and protocols, which as with overly close supervision could end up working against the achievement of high reliability. A culture of high reliability is one in which core norms, values, and rewards are all directed to achieving peak-load performance, safely, all the time, informally as well as formally (T. R. LaPorte, personal communication, 2000). As Rochlin (1993, 21) puts it, "the notion of safe and reliable operation and management must have become so deeply integrated into the culture of the organization that delivery of services and promulgation of safety are held equally as internal goals and objectives: neither can be separated out and 'marginalized' as subordinate to the other, either in normal operations or in emergencies."

RELIABILITY IS NOT FUNGIBLE

Because of the extremely high consequences of error or failure, HROs cannot easily make marginal trade-offs between increasing their services and the reliability with which those services are provided (Rochlin 1993, 16). "Reliability

demands are so intense, and failures so potentially unforgiving, that . . . [m]an-
agers are hardly free to reduce investments and arrive at conclusions about
the marginal impacts on reliability" (Schulman 1993b, 34–35). There is a point
at which the organizations are simply unable to trade reliability for other de-
sired attributes, including money. Money and the like are not interchange-
able with reliability; they cannot substitute for it. High reliability is, formally,
not fungible.

LIMITATIONS ON TRIAL-AND-ERROR LEARNING
(OPERATIONS WITHIN ANTICIPATORY ANALYSIS)

In light of the preceding features, it is not surprising that HROs are very reluctant to allow their primary operations to proceed in a conventional trial-
and-error fashion for fear that the first error would be the last trial (Rochlin
1993, 16). They are characterized by "inability or unwillingness to test the
boundaries of reliability (which means that trial-and-error learning modes become secondary and contingent, rather than primary)" (Rochlin 1993, 23). Although HROs do have search and discovery processes, and elaborate ones,
they will not undertake learning and experimentation that expose them to
greater hazards than they already face. They learn by managing within limits
and, if possible, by setting new limits, rather than testing those limits for errors
(T. R. LaPorte, personal communication, 2000).

As Rochlin puts it, HROs "set goals beyond the boundaries of present performance, while seeking actively to avoid testing the boundaries of error"
(Rochlin 1993, 14). Trial-and-error learning does occur, but this is done outside primary operations, through advanced modeling, simulations, and in
other anticipatory ways that avoid testing the boundary between system continuance and collapse.

FLEXIBLE AUTHORITY PATTERNS UNDER EMERGENCY

HROs "structur[e] themselves to quickly move from completely centralized
decisionmaking and hierarchy during periods of relative calm to completely
decentralized and flat decision structures during 'hot times'" (Mannarelli,
Roberts, and Bea 1996, 84). These organizations have a "flexible delegation of

authority and structure under stress (particularly in crises and emergency situations)" (Rochlin 1996, 56), in which "other, more collegial, patterns of authority relationships emerge as the tempo of operations increases" (LaPorte 1996, 64). When this ability to rapidly decentralize authority under stress is combined with the persistent drive to maintain flexibility and high levels of competence in an HRO, the management emphasis is to work in teams based on trust and mutual respect (T. R. LaPorte, personal communication, 2000). In this way, emergencies can be dealt with by the person on the spot, whose judgment is trusted by other members of the team who work together in these and less charged situations.

POSITIVE, DESIGN-BASED REDUNDANCY TO ENSURE STABILITY OF INPUTS AND OUTPUTS

Last, but certainly not least, HROs are characterized by the multiple ways in which they respond to a given emergency, including having back-up resources and fallback strategies, in order to ensure stability of HROs' inputs and outputs (on the importance of redundancy in systems, see Landau 1969 and Lerner 1986). This positive redundancy is distinguished from other terms, such as *redundancy*, which often have the connotation of "excess capacity." Positive redundancy for high reliability can be designed in several ways:

1. Functional processes are designed so that there are often parallel or overlapping activities that can provide backup in the case of overload or unit breakdown and operation recombination in the face of surprise.

2. Operators and first-line supervisors are trained for multiple jobs including systematic rotation to ensure a wide range of skills and experience redundancy.

It is best to think of the eleven features this way: when there was high reliable service provision—at least during the late 1980s—these features were also generally present. High reliability researchers have been very insistent on maintaining that there are no recipes or formulae for high reliability and, certainly, no guarantees that even if the features were all present, one would find stability of inputs and outputs and highly reliable service provision as a result.

NOTES

CHAPTER I

1. An online search for this term found only one prior reference (Guy 1990), with an entirely different meaning of *management*.

2. Initial findings regarding the effect of the California electricity crisis on grid and service reliability can be found in Roe and others 2002; Roe and others 2005; Schulman and others 2004; de Bruijne and others 2006; and van Eeten and others (2006). Appendix 1 details our research methods, interviews, and sources.

3. For the parallel in regulation, see Mendeloff 1988.

CHAPTER 2

1. This chapter, in particular its description of the CAISO control room in the early years, owes much to the work and writing of Mark de Bruijne (2006).

2. The role of academics and consultants in the policy formulation process has been much remarked upon in the energy fiasco. Gerald Lubenow (2001) of the Institute of Governmental Studies at UC Berkeley noted:

> During the five year process that ended on March 31, 1998, with the opening of California's restructured electricity market, academics were very involved. But while UC experts were often relegated to cameo appearances at formal hearings, the heavy lifting was done by highly paid consultants from eastern campuses. Southern California Edison hired Paul Joskow of MIT. San Diego Gas and Electric retained William Hogan of Harvard.

There were two competing proposals. The PoolCo proposal was, according to Diane Hawk's doctoral thesis [Hawk 1999], "grounded in neoclassical micro-economic theory. The Direct Access proposal is predicated on extent [*sic*] business practices and applied understanding of how competitive markets work. The underlying principles of Direct Access are derived from the collective experience of a diverse group of market participants." The Direct Access proposal, added Hawk, was presented as a product of business experience rather than any abstract academic theory. The model that was eventually adopted was based on a variant of the PoolCo proposal designed by William Hogan and Sally Hunt of National Economic Research Associates.

This time around, the PUC has again relied heavily on paid academic consultants. But instead of looking to the East, they are using the Law and Economic Consulting Group, which is made up largely of Berkeley faculty. At least some Berkeley faculty learned a lesson from the first round of restructuring.

3. If load and generation are not balanced, abnormal system conditions arise that require immediate manual or automatic action to prevent loss of load or power overload, subsequent line and equipment damage, and a cascade of system failures that might result in widespread power outages, even beyond the California grid.

4. Shortfalls in generation capacity did not disappear with the electricity crisis. In November 2005 a senior CAISO official told us, "We've just had fifteen blown load forecasts. The models aren't much use, as temps have been 10 to 15 degrees over the forecast. I can't be left to eight o'clock in the morning to find this out. We can't fire up the short-term extra capacity needed [as a result of this temperature increase] on that short notice."

5. California has not been alone in encountering problems with electricity deregulation. The *Wall Street Journal* summarized the findings of a joint U.S.-Canadian analysis of the August 2003 Northeast blackout: "The blackout resulted from an unfortunate accumulation of problems that have been mounting since wholesale power markets were deregulated in 1992, allowing the buying and selling of bulk electricity at free market prices across swaths of the country" (Smith 2003, A6).

6. For more on confusion and its challenges to crisis management, see Boin, 't Hart, Stern, and Sundelius 2005.

7. Normal days are important because they contain "quiet time," as one shift manager described it. It is welcomed downtime in which operators recharge themselves, work through the backlog of things to do, or prepare for the next peak day. But quiet time also serves other functions. First, the periods slow time down, allowing regularity and the everyday back into the control room. This means time to think things through, not just to catch up. Second, what is background on peak days moves to foreground on quiet ones. Noise subsides; stillness can take over. Instead of facing outward onto the floor to talk or communicate with others, operators swivel round and face the computer screens or documents next to them.

CHAPTER 4

1. This title is taken from LaPorte and Consolini 1991.

2. At the time of the 2003 Northeast Blackout, "the nation's 150,000 miles of heavy-duty power-transmission lines [were] overseen by 130 different controllers, with uncertain lines of authority, making communication choppy and complicating the task of moving

electricity long distances" (Smith, Murray, and Fialka 2003, A1). Another commentator estimated that at the time of the 2003 blackout, the entire grid "included 6,000 power plants run by 3,000 utilities overseen by 142 regional control rooms" (Homer-Dixon 2005, A27).

3. For a brief comparison of NAT and HRO approaches to the August 2003 blackout, see Revkin 2003.

4. This parallels the "law of requisite variety" formulated by Ross Ashby (1952) and more recently extended into the world of organizations by Karl Weick (1995).

5. Middle-level managers are also frequently neglected professionals in the corporate sector (Hymowitz 2005).

6. The analogous argument was famously made by the mathematician Gödel concerning the impossibility of developing a complex set of mathematical propositions and axioms that could be simultaneously complete and consistent.

CHAPTER 5

1. Luck—whether good or bad—and near misses often go together in a control room. A senior grid operations engineer at PG&E described the following close call to us in 2001,

> [I]t's a hot day in May 1987, and we had three pumps out. We nearly caused a voltage collapse all over the western grid. Everything was going up and down, we were trying to get power from all the nuclear units in the western grid. Life flashed before our eyes. And then the gen dispatcher did intuitively the right thing. He said, "Shut one pump down." How he figured that, I still don't understand. It was something we had never seen before. We had no procedures. . . . We went back and looked at it, and the planner said, "Oh yeah, you should never have been running three pumps," and we said, "Where did you get that from?" So we started writing new procedures.

What operators call luck, in other words, can at times be real-time improvisation, though understandably involuntary and with far wider implications in the just-for-now performance mode.

2. The edits have been few and for readability's sake. We have also inserted reference citations and explanatory notes.

3. The vice president of operations is

> responsible for managing system reliability within the control area of the California Independent System Operator (California ISO). Overseeing the organization's Grid Operations, Scheduling and Engineering areas, [the vice president] is responsible for coordination of all transmission and generation maintenance and operations activities. His duties include maintaining the Minimum Operating Reliability Criteria (MORC) established by the North American Electric Reliability Council (NERC, which became at the start of 2007 the North American Electric Reliability Corporation) and the Western Electricity Coordinating Council (WECC [name changed from WSCC after 2001]). He is also responsible for reporting on operations in general and investigating and reporting on system disturbances that occur in the control area. (www.caiso.com/docs/2005/10/12/2005101220305027021.html, accessed October 16, 2005).

4. The new title, "reliability coordinator," replaced the earlier "WSCC security coordinator," while, as noted elsewhere, the acronym WSCC has been replaced by WECC (the Western Electricity Coordinating Council).

5. More specifically, the Real-Time Market Applications (RTMA) system automatically instructs generators to increase or decrease power outputs based on an optimizing model incorporating current market price, forecasted load, line conditions, and a variety of other grid variables. RTMA replaced the manual inc-ing and dec-ing of the BEEPer.

6. "A contingency changes everything" is how one gen dispatcher described the crux of high reliability management.

7. On the importance of mindfulness in high reliability performance, see Weick and Sutcliffe 2001.

8. On the importance to high reliability organizations of the continuous search for improvement, see Appendix 2 and van Eeten and Roe 2002.

9. A CAISO transmission planner told us in 2001: "Engineers always like to do new things, so that is also going on, for example, innovations in static condensers. What we see more is that they are cutting all the slack out of the system, using RASs. . . . "

CHAPTER 6

1. In the words of a senior state government official we interviewed in 2001: "So we deregulated and everybody was happy for three years, and then the missing pieces made the thing blow up. . . . What went wrong in the California deregulation is that different parts of the responsibilities were not located anywhere, such as the responsibility to maintain adequacy and safety." Restructuring left us with an underdesigned system. "All of this happened because the network left a bunch of things unassigned," he continued.

2. Including one of the authors (Roe).

3. In November 2005, a senior control room official reported about the realignment, "we're out on a limb and only getting by with what we have. . . . we're doing our job but we've lost ten to twelve FTE [full-time equivalent employees] here as well. I'm just spraying dog shit right now."

4. This turned out, upon later probing by control room gen dispatchers, to be available only if a state of emergency was declared first by CAISO.

5. On September 28, 2005, when approaching a load peak, a CAISO load dispatcher phoned an operator in the control room of the Los Angeles Department of Water and Power, a city-owned utility, to see if the Department could provide any extra energy for the hour ahead. The LADWP operator responded, "the first megawatt will cost you $186 million" (referring to an unresolved payment dispute from the 2001 energy crisis). The brief conversation ended with a summary brush-off from the LADWP operator: "I have nothing available for you; we don't do business with you."

CHAPTER 7

1. "Complacency really bothers me," one shift supervisor reported to us in 2004. "But complacency comes in two different ways. . . . First, it can happen when you think you know it all, you're overconfident, you let things go on and on thinking it's OK but in the

end you get hammered. That's a complacency issue. I've seen it a million times. The other reason for complacency is because it's just real slow."

2. This was the hallmark of the scientific management literature of the early twentieth century.

CHAPTER 8

1. For an insightful and detailed analysis of the group dynamics associated with a knowledge base in nuclear power plant settings, see Perrin 2005.

2. Although this book is focused on the networked professionals within CAISO, professionals can be found across the generation, transmission, and distribution units in this critical infrastructure for electricity. Cross-professional links can even be glimpsed in the aftermath of the Enron debacle. In the view of one report, the "disappearance of a huge participant [Enron] might have been expected to have a big impact on U.S. energy markets. Yet the lights stayed on, the gas continued to flow. Because they were professional-to-professional markets, there was no damaging impact on consumers" (*Financial Times* 2002, 15).

3. It may seem odd at first to describe adjusting an observed pattern to an action scenario, but this very process has been described in research on "sensemaking" (Weick 1995) and on "recognition-primed decision-making" (Klein 1998).

4. See also Longstaff 2004 for a related distinction between resilience and anticipatory resistance.

5. Widespread blackouts typically reinforce the priority of high reliability management. As the secretary of energy put it prior to the release of the report on the 2003 Northeast blackout, what "has become quite clear is that interconnected operation [of the nation's electricity grid] . . . has to be very, very well run and effective 24 hours a day, 60 minutes per hour, 60 seconds per minute" (quoted in Sevastopulo and Harding 2003: 2).

6. On the key role of surprise in complex systems, see Demchak 1991.

7. See, for example, *The Economist* 2006 for a discussion of the electrical grid as a complex adaptive system. See also Amin 2001 and Haase 2001.

CHAPTER 9

1. This chapter owes much to an earlier collaboration and writing with our colleague Michel van Eeten.

2. "From my view," a shift supervisor told us in 2004, "the grid is more reliable in summer than in winter. In the winter we were very close every day to a major blackout. Problem is the generators are off-line getting repairs and this makes us more dependent on imports, and we're one contingency away from a blackout."

3. "The blackout of 2003 offers a simple but powerful lesson," writes a lead correspondent from the *Wall Street Journal*. "Markets are a great way to organize economic activity, but they need adult supervision" (Wessel 2003, A1). On the 2005 Energy Policy Act and its reforms, see *The Economist* 2006.

4. A senior PG&E grid operations engineer described to us in 2001 the very close links he had with control room operators: "Our job was to provide real-time understanding of the system, that means looking at the data and talking to the dispatchers. We developed a

strong relationship with them and listened carefully. They told me something, and I assumed that I didn't have to run load flows to see if they were right. I said, if you tell me this happened, I will find a way to model it. . . . The whole thing for me was observe, tell me, model. We started new procedures that sometimes got them [the operators] out of trouble and helped them look at unidentified risks, and that built confidence with them."

CHAPTER 10

1. When we gave our presentation to a meeting of CAISO operations staff on the problems raised in our January 2005 report, the VP-Ops gave the first comment after we finished. "Don't close off on this too soon. Don't rush off," he told his staff. "Let's think about this before we try to solve it."

CHAPTER 11

1. This perception of the "null hypothesis" regarding reliability is detailed in de Bruijne 2006.

2. A CPS2 violation occurs when the Area Control Error (ACE) over a ten-minute period is outside its regulatory bandwidth. High RTMA bias reflects the number of times that the bias introduced into RTMA (the dispatching software) was greater than 400MW (in a positive or negative direction). Bias is used when the generation dispatcher does not trust the load forecast upon which RTMA solutions and dispatches will be based. Mitigations occur when operators make out-of-sequence or out-of-market dispatches to deal with grid congestion—the inability of transmission lines to cover needed energy flows. The number of unscheduled outages reflects the instances when generation capacity goes offline unexpectedly.

3. Reliable information on RTMA, which started in October 2004, was unavailable until November 2004 (thus analysis with RTMA bias begins in November). We excluded variables for which data were not available (for example, daily workarounds to compensate for software glitches). The start of the analysis (July 2004) was chosen because that was when we first suspected the operators' reliability management had become problematic (Chapter 5), whereas the end (December 31, 2006) was the last full month for which we had data.

4. The use of regression analysis contrasts with how economists and engineers model the electricity sector. For them, the electricity system has so many interacting variables and feedback loops that its modeling requires operations research, dynamic modeling, and econometrics. The operators we interviewed know how complex and dynamic the electric grid is. But their reliability focus is localized and time-defined, when controlling for an input is meant to have a visible and determinable effect on an output. This is what linear regression analysis measures. Methodologically, we find ourselves in a position similar to that of Cyert and March in their *Behavioral Theory of the Firm* (1963). They too started by identifying the rules of thumb and heuristics producers actually followed rather than what optimal process theory said should be followed.

5. In the second phase of research, a new data source enabled us to push the date back to November 4, 2004.

6. We also found another period in which operators moved to their CPS2 edge. What was called the "C1 Network Model Changeover," including changes involving an adjacent

control area, took effect December 1, 2005. For this changeover subperiod, the adjusted R^2 rose to 0.44.

7. The CPS2 standard, established by the North American Electric Reliability Council (NERC, now Corporation), requires CAISO to be within the ACE bandwidth 90 percent of its operating time. This means that it can effectively average no more than fourteen CPS2 violations per day during any month.

8. Changing the daily period to sixty-five days or ninety days did not substantially change the configuration of the figures that appear in the text. The actual moving range period was fifty-one days, to coincide with the RTMA introduction period for which we originally had data, namely, November 11 through December 31, 2004 (inclusive).

9. Because the CPS2 edge has RTMA as an input variable, the study period was limited to those dates for which high biasing information was available. A dataset of moving range R^2s was created. Only those R^2s whose respective F-statistics were statistically significant and whose regression coefficients were statistically significant for two or more input variables were considered. R^2s were then ranked from highest to lowest, separated into equal thirds, with the highest and lowest cohort considered. Baseline means and standard deviations were calculated for the three input variables and the output variable. The mean and standard deviations for each variable were also calculated for each range subperiod and then filtered into two mutually exclusive categories: those range subperiods with values that were over the baseline values and those that were at or less than the baseline values. This was done for each input variable and the output variable. In the case of mixed results—for example, not all three input variables have means and standard deviations that exceed their baseline counterparts—the filtering rule became the following: If two or more inputs have values that are over the respective baseline, then they are grouped in the "over baseline" cell. Both means and standard deviations must be either over or at/below to qualify for selection into one of the typology's internal cells.

10. Both cohorts had different subperiods that contained the same days; there were forty-three overlapping days out of the three hundred days across the two cohorts. The dummy variable regression coefficient remained statistically significant, whether the overlapping days were coded 1 or 0. Doing so did not change the statistical significance of the other coefficients.

11. The shape of the subperiod spike curve in Figure 11.4 does not appear to be sensitive to adding or subtracting a day or so at either end. When operators were shown the baseline curve of average hourly CPS2 violations over the entire study period, they said it matched what they knew: Hour Ending (HE)7 (off peak to on peak), HE18 (around actual peak load), and HE23 (on peak to off peak) are more difficult than other hours of the day in terms of control room operations.

PART III

1. The interagency coordination of California's large water supplies continues to the time of writing. One of us (Roe) has followed the weekly conference calls of the state's Data Assessment Team involving operational managers from California's water supply and fish agencies, who help coordinate real-time water transfers in light of state and federal environmental mandates during a sensitive portion of the year.

2. On the importance of both electricity and telecommunications to other critical infrastructures, see National Research Council 2002, Figure 10.4.

CHAPTER 12

1. When asked about the educational requirements of residential metering for its end users, an economist and proponent of metering told us, "That is an interesting question. I haven't really thought about it." At a conference on demand response, the engineers said residential metering should be as easy as using the ATM. Of course, when one ATM is not working you go to another. When asked how the user of the residential metering device could do that, if the device was not working, they replied, "Buy another one."

CHAPTER 13

1. The Panel report, along with reports on other BP mishaps, has provoked the new head of BP "to say that his first priority will be to focus 'like a laser' on safe and reliable operations" (Crooks 2007, 3).

2. Although air traffic control reflects the public sector, it is important to underscore that safety is core to private sector critical infrastructures as well. Examples abound. At the time of writing, BP, the petroleum giant, has been criticized for major mishaps in Alaska as well (McNulty 2006, 2).

3. A control room employee filed suit against CAISO arguing that because he was not formally classified as a professional employee, he was entitled to overtime pay for more than eight hours in his shift. A judge agreed, and as part of the settlement control room employees became hourly wage employees, with among other requirements mandatory lunch breaks away from the control room. Besides the lowering of morale, in one instance an experienced gen dispatcher was unavailable during an emergency because he had to take a half-hour break away from the work site.

4. See, for example, Milward and Proven (2000, 253) on the importance of stable networks to effective performance in human service systems.

5. As one gen dispatcher jokingly asserted, "We're not really the CAISO control area, we're the CAISO request area."

6. In early interviews at the CPUC we repeatedly encountered skepticism about CAISO's competence in pursuing the cheapest market solutions to its power needs. This is matched by a skepticism within CAISO about how well CPUC officials really understand what it takes to manage a statewide grid. The relations between CAISO and CPUC have improved since 2001. One senior control room official told us in the latest round of interviews, "We have had a more collaborative effort with the PUC and Energy Commission and the governor's office, and they routinely call us to figure out what's the right thing to do. They actually want this to work." A CAISO staff member recently looked back to 2001 and having to work with an earlier CPUC chair as "like having Hezbollah in the room."

7. We should be clear about what we are recommending here. A professional identity that promotes cross-communication and the sharing of skills and knowledge is quite different from a formal certification as "professionals." Ironically certification grounded in formal curricula and credentials could place reliability professionals in the same position as

macro-designers, operating from principles more than from the mix of formal and tacit knowledge that defines their domain of competence. A discussion of this risk can be found in Larsen 1984 (page 46).

8. In late 2005, a senior control room official concluded about the CAISO realignment, "I figured that just in the control room we lost 250 person years of training in the 10 to 12 people we lost. The people we laid off had twenty years experience, while the people who are here now have ten to fifteen years experience. We made the gamble to respond sooner than later to the crisis in the industry [that is, that experienced people are retiring throughout the electricity sector], but in doing so, we opened a window of vulnerability on the floor, but that has been compensated we hope by better teams."

9. "Economics has no body of professional ethics, and never has," writes one economist (DeMartino 2005, 89). "Were one to phone the most prestigious economics programs in the country—even those that train practitioners who will take up leading posts in government and multinational agencies and whose work will shape the life circumstances of millions—and ask to be connected to their professional economics ethicist, one will encounter only confusion." DeMartino goes on to draft an "Economist's Oath" (after the Hippocratic Oath), one part of which reads, "You [the aspiring economist] will teach those you instruct and with whom you work of the vagaries of the practice of economics, alert them to the dangers of economic experimentation, and, to the best of your ability, help them to anticipate and prepare for unintended consequences." Amen.

APPENDIX 2

1. Material in this appendix appeared in slightly altered form in van Eeten and Roe 2002, Chapter 3, "Recasting the Paradox Through a Framework of Ecosystem Management Regimes." Used by permission of Oxford University Press.

REFERENCES

Alaywan, Z. 2000. "Evolution of the California Independent System Operator Markets." *The Electricity Journal* 13(6): 70–83.

Allenby, B., and J. Fink. 2005, August 12. "Toward Inherently Secure and Resilient Societies." *Science* 309: 1,034–1,036.

Amin, M. 2001. "Complex Interactive Networks Workshop." Washington, DC: EPRI Grid Operations and Planning.

Apt, J., L. Lave, S. Talukdar, M. G. Morgan, and M. Ilic. 2004. "Electrical Blackouts: A Systemic Problem." *Issues in Science and Technology* 20(4): 55–61.

Ashby, R. 1952. *Design for a Brain.* London: Chapman and Hall.

Auerswald, P., L. Branscomb, T. M. LaPorte, and E. Michel-Kerjan (Eds.). 2006. *Seeds of Disaster, Roots of Response: How Private Action Can Reduce Public Vulnerability.* New York: Cambridge University Press.

Barbose, G., C. Goldman, and B. Neenan. 2004. *A Survey of Utility Experience with Real Time Pricing.* LBNL-54238. Berkeley, CA: Lawrence Berkeley National Laboratory.

Bardach, E. 2005. *A Practical Guide for Policy Analysis,* 2nd ed. Washington, DC: CQ Press.

Bardach, E., and R. Kagan. 1982. *Going by the Book: The Problem of Regulatory Unreasonableness.* Philadelphia: Temple University Press.

Beamish, T. D. 2002. *Silent Spill.* Cambridge, MA: The MIT Press.

Bennis, W., and B. Nanus. 1997. *Leaders: Strategies for Taking Charge.* New York: HarperCollins.

Blumstein, C., L. S. Friedman, and R. Green. 2002. "The History of Electricity Restructuring in California." *Journal of Industry, Competition and Trade* 2(1–2): 9–38.

Boin, A. 2004. "The Early Years of Public Institutions: A Research Agenda." International Conference on the Birth of Institutions, Leiden, Netherlands, June 10–12.

Boin, A., and A. McConnell. 2007. "Preparing for Critical Infrastructure Breakdowns: The Limits of Crisis Management and the Need for Resilience." *Journal of Contingencies and Crisis Management* 15(1): 50–59.

Boin, A., P. 't Hart, E. Stern, and B. Sundelius. 2005. *The Politics of Crisis Management: Public Leadership Under Pressure.* Cambridge, UK: Cambridge University Press.

Borenstein, S. 2004. *The Long-Run Effects of Real-Time Electricity Pricing.* CSEM WP 133. Berkeley, CA: University of California Energy Institute, Center for the Study of Energy Markets.

Bosk, C. 2003. *Forgive and Remember: Managing Medical Failure.* Chicago: University of Chicago Press.

Bovens, M., and P. 't Hart. 1996. *Understanding Policy Fiascos.* New Brunswick, NJ: Transaction.

BP U.S. Refineries Independent Safety Review Panel. 2007. *Report of the BP U.S. Refineries Independent Safety Review Panel.* Houston: British Petroleum.

Brunner, R., and A. Willard. 2003. "Professional Insecurities: A Guide to Understanding and Career Management." *Policy Sciences* 36: 3–36.

California Public Utilities Commission. 1993. *California's Electric Services Industry: Perspectives on the Past, Strategies for the Future.* San Francisco: Division of Strategic Planning.

California State Auditor. 2001. *Energy Deregulation: The State's Energy Balance Remains Uncertain but Could Improve with Changes to Its Energy Programs and Generation and Transmission Siting.* 2000-134.2. Sacramento, CA: California State Auditor.

Campbell, D. T. 1988. *Methodology and Epistemology for Social Sciences: Selected Papers.* E. S. Overman (Ed.). Chicago: University of Chicago Press.

Carroll, J. S. 2003. "Knowledge Management in High-Hazard Industries: Accident Precursors as Practice". In J. R. Phimister, et al. (Eds.), *Accident Precursor Analysis and Management.* Washington, DC: National Academy of Engineering.

Comfort, L. 1999. *Shared Risk: Complex Systems in Seismic Response.* New York: Pergamon.

Conference on Critical Infrastructure Protection. 2007, April 18–19. *Critical Infrastructure Protection: Issues for Resilient Design.* The Hague, The Netherlands.

Congressional Budget Office. 2001. *Causes and Lessons of the California Electricity Crisis.* Washington, DC: Congress of the United States.

Crooks, E. 2007. "Laid Low by Scandal." Energy Special Report, *Financial Times,* June 19, 3.

Cyert, R., and J. G. March. 1963. *A Behavioral Theory of the Firm.* Englewood Cliffs, NJ: Prentice-Hall.

de Bruijne, M. 2006. *Networked Reliability: Institutional Fragmentation and the Reliability of Service Provision in Critical Infrastructures.* Delft, The Netherlands: Delft University of Technology.

de Bruijne, M., and M. van Eeten. 2007. "Systems That Should Have Failed: Critical Infrastructure Protection in an Institutionally Fragmented Environment." *Journal of Contingencies and Crisis Management* 15(1): 18–29.

de Bruijne, M., M.J.G. van Eeten, E. Roe, and P. Schulman. 2006. "Assuring High Reliability of Service Provision in Critical Infrastructures." *International Journal of Critical Infrastructures* 2(2–3): 231–246.

DeMartino, G. 2005, July-August. "A Professional Ethics Code for Economists." *Challenge* 48(4): 88–104.

Demchak, C. 1991. *Military Organizations, Complex Machines: Modernization in the U.S. Armed Services.* Ithaca, NY: Cornell University Press.

Dillon, M., and C. Wright (Eds.). 2005. *Complexity, Networks, and Resilience: Interdependence and Security in the 21st Century.* Chatham, NJ: Chatham House.

Duane, T. P. 2002. "Regulation's Rationale: Learning from the California Energy Crisis." *Yale Journal on Regulation* 19(2): 471–540.

Duncan, A. J. 1986. *Quality Control and Industrial Statistics.* Homewood, IL: Irwin.

The Economist. 2006. "Electricity Supply: More Heat than Light," July 29: 30.

Evan, W. M., and M. Manion. 2002. *Minding the Machines: Preventing Technological Disasters.* Upper Saddle River, NJ: Prentice-Hall.

Farrell, A. E., L. B. Lave, and G. Morgan. 2002. "Bolstering the Security of the Electric Power System." *Issues in Science and Technology* 18(3): 49–56.

Farson, R., and R. Keyes. 2002, August. "The Failure-Tolerant Leader." *Harvard Business Review:* 64–71.

Financial Times. 2002. "A Fresh Look at Rules for Energy and Finance," February 19: 15.

Financial Times. 2005. "Special Report on Business Continuity," June 27.

Frederickson, H. G., and T. LaPorte. 2002, September. "Airport Security, High Reliability, and the Problem of Rationality." *Public Administration Review* 62 (Special Issue): 33–43.

Gawande, A. A. 2003. *Complications: A Surgeon's Notes on an Imperfect Science.* New York: Picador USA.

Glicken, M. D. 2006. *Learning from Resilient People.* Thousand Oaks, CA: Sage.

Gourevitch, P. A., and J. Shinn. 2005. *Political Power and Corporate Control.* Princeton, NJ: Princeton University Press.

Groopman, J. 2007. *How Doctors Think.* New York: Houghton Mifflin.

Guy, M. E. 1990. "High-Reliability Management." *Public Productivity & Management Review,* 13(4): 301-313.

Haase, P. 2001, Spring. "Of Horseshoe Nails and Kingdoms: Control of Complex Interactive Networks and Systems." *EPRI Journal:* 1–10.

Hammond, K. 2000. *Judgments Under Stress.* New York: Oxford University Press.

Harrison, M., and A. Shirom. 1999. *Organizational Diagnosis and Assessment: Bridging Theory and Practice.* Thousand Oaks, CA: Sage.

Hauer, J. F., and J. E. Dagle. 1999. *Grid of the Future: White Paper on Review of Recent Reliability Issues and System Events.* Consortium for Electric Reliability Technology Solutions, PNNL-13150. Richland, WA: Pacific Northwest National Laboratory.

Hawk, D. V. 1999. *Disconnect: A Transaction Cost Analysis of California Electric Power Industry Restructuring.* Ph.D. Dissertation. Berkeley, CA: University of California.

Hey, D. L., and N. S. Philippi. 1994. *Reinventing Flood Control Strategy.* Chicago: The Wetlands Initiative.

Hill, M., and P. Hupe. 2002. *Implementing Public Policy: Governance in Theory and in Practice.* London: Sage.

Hodgkinson, G., and P. Sparrow. 2002. *The Competent Organization: A Psychological Analysis of the Strategic Management Process.* Buckingham, UK: Open University Press.

Homer-Dixon, T. 2005. "Caught Up in Our Own Connections." Op-ed, *New York Times,* August 13: A27.

Hutter, B., and M. Power. 2005. "Organizational Encounters with Risk: An Introduction." In B. Hutter and M. Power (Eds.), *Organizational Encounters with Risk.* Cambridge, UK: Cambridge University Press.

Hymowitz, C. 2005. "Middle Managers Are Unsung Heroes on Corporate Stage." *Wall Street Journal,* September 19: B1.

Imperial, M. 2005. "Using Collaboration as a Governance Strategy: Lessons from Six Watershed Management Programs." *Administration and Society* 37: 281–320.

Institute of Medicine. 2000. *To Err Is Human: Building a Safer Health System.* Washington, DC: The National Academy Press.

Issing, O., and V. Gaspar (with O. Tristani, and D. Vestin). 2005. *Imperfect Knowledge and Monetary Policy.* The Stone Lectures in Economics. Cambridge, UK: Cambridge University Press.

Ives, A. R., and S. R. Carpenter. 2007, July 6. "Stability and Diversity in Ecosystems." *Science* 307: 58–62.

Joskow, P. L. 2001a. "California's Electricity Crisis." NBER Working Paper Series, Working Paper 8442. Accessed online at www.nber.org/papers/w8442.

Joskow, P. L. 2001b. "U.S. Energy Policy During the 1990s." Paper presented for conference "American Economic Policy During the 1990s," John F. Kennedy School of Government, Harvard University, June 27–30.

Juran, J. M. 1986. "Early SQC: A Historical Supplement." *Quality Progress* 30: 73–81.

Kahn, M., and L. Lynch. 2000. *California's Electricity Options and Challenges.* Sacramento, CA: Electricity Oversight Board and California Public Utilities Commission.

Kaplan, F. 1991. *The Wizards of Armageddon.* Stanford, CA: Stanford University Press.

Kay, J. 2006. "Why the Key to Carbon Trading Is to Keep It Simple." *Financial Times,* May 9: 15.

Kelman, S. 2006. "Book Review Essay: 9/11 and the Challenges of Public Management." *Administrative Science Quarterly* 51: 129–142.

Klein, G. 1998. *Sources of Power.* Cambridge, MA: The MIT Press.

Klein, K., J. Ziegert, A. Knight, and Y. Xiao. 2006. "Dynamic Delegation: Shared, Hierarchical, and Deindividualized Leadership in Extreme Action Teams." *Administrative Science Quarterly* 51: 590–621.

Knapp, R. A., K. R. Matthews, and O. Sarnelle. 2001. "Resistance and Resilience of Alpine Lake Fauna Assemblages to Fish Introductions." *Ecological Monographs* 71: 401–421.

Korten, D. 1980. "Community Organization and Rural Development: A Learning Process Approach." *Public Administration Review* 40(5): 480–511.

Landau, M. 1969, July-August. "Redundancy, Rationality, and the Problem of Duplication and Overlap." *Public Administration Review* 29: 346–358.

Langer, E. 1989. *Mindfulness.* New York: Addison-Wesley.

LaPorte, T. R. 1993. "'Organization and Safety in Large Scale Technical Organizations: Lessons from High Reliability Organizations Research and Task Force on 'Institutional Trustworthiness.'" Paper prepared for a seminar "Man-Technology-Organization in Nuclear Power Plants," Finnish Centre for Radiation and Nuclear Safety, Technical Research Centre of Finland, Olkiluoto, Finland. June 14–15.

LaPorte, T. R. 1994. "A Strawman Speaks Up: Comments on Limits of Safety." *Journal of Contingencies and Crisis Management* 2(4): 207–211.

LaPorte, T. R. 1996. "High Reliability Organizations: Unlikely, Demanding and at Risk." *Journal of Contingencies and Crisis Management* 4(2): 60–71.

LaPorte, T. R., and P. Consolini. (1991). "Working in Practice but Not in Theory: Theoretical Challenges of High Reliability Organizations." *Public Administration Research and Theory* 1(1): 19–47.

LaPorte, T. R., and T. Lascher. 1988. *Cold Turkeys and Task Forces: Pursuing High Reliability in California's Central Valley.* Working Paper 88–25, Institute of Governmental Studies. Berkeley, CA: University of California.

Larson, M. S. 1984. "The Production of Expertise and the Constitution of Expert Power." In T. L. Haskell (Ed.), *The Authority of Experts,* 28–80. Bloomington, IN: Indiana University Press.

Lasswell, H. 1951. "The Policy Orientation." In D. Lerner and H. Lasswell (Eds.), *The Policy Sciences: Recent Development in Scope and Method.* Stanford, CA: Stanford University Press.

Lawrence Livermore National Laboratory. 1998, January-February. "Making Information Safe." *Science & Technology Review.* Livermore, CA: University of California.

Leaf, C. 2004. "Why We Are Not Winning the War on Cancer (and How to Win It)." *Fortune* 149(6): 76ff.

Lee, S. T. 2001. *Analysis of the 2000–2001 California Power Crisis: Implications for Grid Operations and Planning.* Palo Alto, CA: Electric Power Research Institute.

Leonard, M., A. Frankel, and T. Simmonds (with K. Vega). 2004. *Achieving Safe and Reliable Healthcare: Strategies and Solutions.* ACHE Management Series. Chicago: Health Administration Press.

Lerner, A. W. 1986. "There Is More Than One Way to Be Redundant: A Comparison of Alternatives for the Design and Use of Redundancy in Organizations." *Administration & Society* 18(3): 334–359.

Lerner, D., and H. Lasswell (Eds.). 1951. *The Policy Sciences: Recent Development in Scope and Method.* Stanford, CA: Stanford University Press.

Longstaff, P. H. 2004. "Security in Complex, Unpredictable Systems: The Case for Resilience Planning." New York: Harvard University, Center for Information Policy Research, and Syracuse University, S.I. Newhouse School of Public Communications.

Lubenow, G. 2001. "Take My Advice, Please." *Public Affairs Report* 42(1). Berkeley, CA: University of California, Institute of Governmental Studies.

Majone, G. 1978. "Technology Assessment in a Dialectic Key." *Public Administration Review* 38(1): 52–58.

Mannarelli, T., K. Roberts, and R. Bea. 1996. "Learning How Organizations Mitigate Risk." *Journal of Contingencies and Crisis Management* 4(2): 83–92.

Marquand, D. 2004. *Decline of the Public.* Cambridge, UK: Polity Press.

Marris, P., and M. Rein. 1969. *Dilemmas of Social Reform.* New York: Atherton Press.

Maruyama, M. 1963. "The Second Cybernetics: Deviance-Amplifying Mutual Causal Processes." *American Scientist* 51(2): 164–179.

McNulty, S. 2006. "All Symptoms of a Failed Safety Culture." Special Report: Energy. *Financial Times,* October 23: 2.

Mendeloff, J. 1988. *The Dilemma of Toxic Substance Regulation: How Overregulation Leads to Underregulation.* Cambridge, MA: The MIT Press.

Mensah-Bonsu, C., and S. Oren. 2001. *California Electricity Market Crisis: Causes, Remedies and Prevention.* Folsom, CA: California Independent System Operator (CAISO).

Meshkati, N. 1991. "Human Factors in Large-Scale Technological Systems' Accidents: Three Mile Island, Bhopal, Chernobyl." *Industrial Crisis Quarterly* 5(2): 133–154.

Michael, D. 1973. *On Learning to Plan and Planning to Learn.* San Francisco: Jossey-Bass.

Milward, H. B., and K. G. Provan. 2000. "How Networks Are Governed." In C. R. Heinrich and L. E. Lynn (Eds.), *Governance and Performance: New Perspectives,* 238–262. Washington, DC: Georgetown University Press.

Mintzberg, H. 1979. *The Structuring of Organizations.* Englewood Cliffs, NJ: Prentice-Hall.

National Academy of Engineering. 2003. *Accident Precursor Analysis and Management: Reducing Technological Risk Through Diligence.* Washington, DC: National Academy Press.

National Research Council. 2002. *Making the Nation Safer: The Role of Science and Technology in Countering Terrorism.* Washington, DC: Committee on Science and Technology for Countering Terrorism, National Academy Press.

Norman, D. 2002. *The Design of Everyday Things.* New York: Basic Books.

Ott, E. R., and E. G. Schilling. 1990. *Process Quality Control.* New York: McGraw-Hill.

Paret, P. 1986. "Clausewitz." In P. Paret (Ed., with collaboration of G. Craig and F. Gilbert), *Makers of Modern Strategy: From Machiavelli to the Nuclear Age,* 2nd ed. Princeton, NJ: Princeton University Press.

Perrin, C. 2005. *Shouldering Risks.* Princeton, NJ: Princeton University Press.

Perrow, C. 1979. *Complex Organizations: A Critical Essay.* New York: Wadsworth.

Perrow, C. 1983. "The Organizational Context of Human Factors Engineering." *Administrative Science Quarterly* 28: 521–541.

Perrow, C. 1994. "Limits of Safety." *Journal of Contingencies and Crisis Management* 2(4): 212–220.

Perrow, C. 1999 [1984]. *Normal Accidents.* Princeton, NJ: Princeton University Press.

Plender, J. 2002. "What Price Virtue? The Anglo-U.S. Model Has Gone Off the Rails Because the Penalties and Rewards Are Skewed." *Financial Times,* December 2: 13.

Pool, R. 1997. *Beyond Engineering: How Society Shapes Technology.* New York: Oxford University Press.

Reason, J. 1972. *Human Error.* Cambridge, UK: Cambridge University Press.

Regalado, A., and G. Fields. 2003. "Blackout a Reminder of Grid's Vulnerability to Terror." *Wall Street Journal,* August 15: A3.

Revkin, A. 2003. "The Thrill of Danger, the Agony of Disaster." *New York Times Week in Review,* August 31, Sec. 4: 1, 5.

Rijpma, J. A. 1997. "Complexity, Tight Coupling and Reliability: Connecting Normal Accidents Theory and High Reliability Theory." *Journal of Contingencies and Crisis Management* 5(1): 15–23.

Roberts, K. H. 1988. *Some Characteristics of High Reliability Organizations.* Berkeley, CA: University of California, School of Business Administration.

Roberts, K. 1990. "Some Characteristics of One Type of High Reliability Organisation." *Organisation Science* 1(2): 160–176.

Roberts, K. (Ed.). 1993. *New Challenges to Understanding Organizations.* New York: Macmillan.

Rochlin, G. 1993. "Defining 'High Reliability' Organizations in Practice: A Taxonomic Prologue." In K. Roberts (Ed.), *New Challenges to Understanding Organizations,* 11–32. New York: Macmillan.

Rochlin, G. 1996. "Reliable Organizations: Present Research and Future Directions." *Journal of Contingencies and Crisis Management* 4(2): 55–59.

Rochlin, G., and A. von Meier. 1994. "Nuclear Power Operations: A Cross Cultural Perspective. *Annual Review of Energy and Environment* 19: 153–187.

Rochlin, G., T. R. LaPorte, and K. Roberts. 1987. "The Self-Designing High-Reliability Organization: Aircraft Carrier Flight Operations at Sea." *Naval War College Review* 40(4): 76–90.

Roe, E. 1994. *Narrative Policy Analysis.* Durham, NC: Duke University Press.

Roe, E. 1999. *Except-Africa: Remaking Development, Rethinking Power.* New Brunswick, NJ: Transaction.

Roe, E., M. van Eeten, P. Schulman, and M. de Bruijne. 2002. *California's Electricity Restructuring: The Challenge of Providing Service and Grid Reliability.* EPRI Report 1007388, prepared for the California Energy Commission and Lawrence Berkeley National Laboratory. Palo Alto, CA: Electric Power Research Institute.

Roe, E., P. Schulman, M.J.G. van Eeten, and M. de Bruijne. 2005. "High Reliability Bandwidth Management in Large Technical Systems: Findings and Implications of Two Case Studies." *Journal of Public Administration Research and Theory* 15(2): 263–280.

Rosenthal, M. M., and K. M. Sutcliffe (Eds.). 2002. *Medical Error.* San Francisco: Jossey-Bass.

Ryle, G. 1949. *The Concept of Mind.* London: Hutchinson's University Library.

Sagan, S. 1993. *The Limits of Safety.* Princeton, NJ: Princeton University Press.

Salamon, L. M. 2002. "The New Governance and the Tools of Public Action." In L. Salamon and O. Elliott (Eds.), *The Tools of Government.* New York: Oxford University Press.

Salvendy, G. 1997. *Handbook of Human Factors and Ergonomics.* New York: Wiley.

Sanne, J. M. 2000. *Creating Safety in Air Traffic Control.* Lund, Sweden: Arkiv Forlag.

Savas, E. S. 1997. *Privatization and Public/Private Partnerships.* Washington, DC: CQ Press.

Schein, E. H. 1994. *Organizational and Managerial Culture as a Facilitator or Inhibitor of Organizational Learning.* Available at www.solonline.org/res/wp/10004.html.

Schulman, P. R. 1980. *Large-Scale Policy-Making.* New York: Elsevier.

Schulman, P. R. 1993a. "The Negotiated Order of Organizational Reliability." *Administration & Society* 25(3): 353–372.

Schulman, P. R. 1993b. "The Analysis of High Reliability Organizations: A Comparative Framework." In K. Roberts (Ed.), *New Challenges to Understanding Organizations,* 33–54. New York: Macmillan.

Schulman, P. R. 2002. "Medical Errors: How Reliable Is Reliability Theory?" In M. M. Rosenthal and K. M. Sutcliffe (Eds.), *Medical Error,* San Francisco: Jossey Bass.

Schulman, P. R. 2004. "General Attributes of Safe Organizations." *Quality Safety in Health Care* 13 (Suppl II): ii39–ii44.

Schulman, P. R., and E. Roe. 2006. "Managing for Reliability in an Age of Terrorism." In P. Auerswald, L. Branscomb, T. R. LaPorte, and E. Michel-Kerjan (Eds.), *Seeds of Disaster, Roots of Response: How Private Action Can Reduce Public Vulnerability, 121–134.* Cambridge, UK: Cambridge University Press.

Schulman, P. R., and E. Roe. 2007. "Dilemmas of Design and the Reliability of Critical Infrastructures." *Journal of Contingencies and Crisis Management* 15(1): 42–49.

Schulman, P. R., E. Roe, M. van Eeten, and M. de Bruijne. 2004. "High Reliability and the Management of Critical Infrastructures." *Journal of Contingencies and Crisis Management* 12(1): 14–28.

Scott, J. C. 1998. *Seeing Like a State: How Certain Schemes to Improve the Human Condition Have Failed.* New Haven, CT: Yale University Press.

Selznick, P. 1957. *Leadership in Administration.* New York: Harper & Row.

Senge, P. 1990. *The Fifth Discipline.* New York: Doubleday.

Sevastopulo, D., and J. Harding. 2003. "Upbeat Energy Chief Prepares to Shine Light on Grid Failure." *Financial Times,* November 17: 2.

Sheffi, Y. 2005. *The Resilient Enterprise.* Cambridge, MA: The MIT Press.

Sibbet, D., and L. Lind (Facilitators). 2000. "California ISO 'Congestion Reform Stakeholder Meeting' Folsom, CA. May 10–11 2000." San Francisco.

Smathers, D., and A. Akhil. 2001. *Operating Environment and Functional Requirements for Intelligent Distributed Control in the Electric Power Grid.* Sandia Report. Albuquerque, NM and Livermore, CA: Sandia National Laboratories.

Smith, R. 2003. "Report Sheds Light on Blackout." *Wall Street Journal,* November 20: A6.

Smith, R., S. Murray, and J. Fialka. 2003. "How Unlikely Coalition Scuttled Plan to Remake Electrical Grid." *Wall Street Journal,* November 4: A1, A14.

Starbuck, W., and M. Farjoun (Eds.). 2005. *Organization at the Limit: Lessons from the Columbia Disaster.* Malden, MA: Blackwell.

Stockman, C., M. Piette, and L. ten Hope. 2004. *Market Transformation Lessons Learned from an Automated Demand Response Test in the Summer and Fall of 2003.* LBNL-55110. Sacramento, CA: California Energy Commission.

Sunstein, C. 2007. *Worst-Case Scenarios.* Cambridge, MA: Harvard University Press.

Sweeney, J. L. 2002. *The California Electricity Crisis.* Palo Alto, CA: Hoover Institution Press.

Tayor, J. 2004. "Lay Off." *Wall Street Journal,* July 9: A10.

Thackara, J. 2006. *In the Bubble: Designing in a Complex World.* Cambridge, MA: The MIT Press.

The Structure Group. 2004. *Grid West: RTO Cost Drivers and Considerations.* Houston: The Structure Group.

Tieman, R. 2006. "Reliability Is Part of Brand Protection." Special Report on Risk Management. *Financial Times,* April 25: 2.

Turner, B. A. 1978. *Man Made Disasters.* London: Wykenham.

van Eeten, M.J.G., and E. Roe. 2002. *Ecology, Engineering and Management: Reconciling Ecosystem Rehabilitation and Service Reliability.* New York: Oxford University Press.

van Eeten, M.J.G., E. Roe, P. Schulman, and M. de Bruijne. 2006. "When Failure Is Not an Option: Managing Complex Technologies Under Intensifying Interdependencies." In Robert Verburg et al. (Eds.), *Managing Technology and Innovation: An Introduction,* 306–322. London: Routledge.

Vaughan, D. 1996. *The Challenger Launch Decision: Risky Technology, Culture and Deviance at NASA.* Chicago: University of Chicago Press.

Vaughan, D. 2005. "Organizational Rituals of Risk and Error." In B. Hutter and M. Power (Eds.), *Organizational Encounters with Risk,* 33–66. Cambridge, UK: Cambridge University Press.

von Meier, A. 1999. "Occupational Cultures as a Challenge to Technological Innovation." *IEEE Transactions on Engineering Management* 46(1): 104–114.

Wagenaar, W. 1996. "Profiling Crisis Management." *Journal of Contingencies and Crisis Management* 4(3): 169–174.

Weare, C. 2003. *The California Electricity Crisis: Causes and Policy Options.* San Francisco: Public Policy Institute of California.

Weick, K. E. 1987. "Organizational Culture as a Source of High Reliability." *California Management Review* 29(2): 112–127.

Weick, K. E. 1993. "The Vulnerable System: An Analysis of the Tenerife Air Disaster." In K. Roberts (Ed.), *New Challenges to Understanding Organizations,* 173–197. New York: Macmillan.

Weick, K. 1995. *Sensemaking in Organizations.* Thousand Oaks, CA: Sage.

Weick, K., and K. Roberts. 1993. "Collective Mind in Organizations: Heedful Interrelating on Flight Decks." *Administrative Science Quarterly* 38(3): 357–381.

Weick, K. E., and K. M. Sutcliffe. 2001. *Managing the Unexpected: Assuring High Performance in an Age of Complexity.* San Francisco: Jossey-Bass.

Weick, K. E., K. M. Sutcliffe, and D. Obstfeld. 1999. "Organizing for High Reliability: Processes of Collective Mindfulness." In B. Straw and L. Cummings (Eds.), *Research in Organisational Behaviour,* 81–123. Greenwich, CT: JAI.

Wessel, D. 2003. "A Lesson from the Blackout: Free Markets Also Need Rules." *Wall Street Journal,* August 28: A1, A2.

Wheat, A. 2001, December 24. "The Top 10 Business Stories." *Fortune* 144(13).

Wildavsky, A. 1988. *Searching for Safety.* New Brunswick, NJ: Transaction.

Wilensky, H. 1964. "The Professionalization of Everyone?" *American Journal of Sociology* 70(1): 137–158.

INDEX